Guide to Semiconductor Engineering

Second Edition

Guide to Semiconductor Engineering

Second Edition

Jerzy Ruzyllo
Pennsylvania State University, USA

World Scientific

NEW JERSEY • LONDON • SINGAPORE • BEIJING • SHANGHAI • HONG KONG • TAIPEI • CHENNAI • TOKYO

Published by

World Scientific Publishing Co Pte Ltd
5 Toh Tuck Link, Singapore 596224
USA office: 27 Warren Street, Suite 401-402, Hackensack, NJ 07601
UK office: 57 Shelton Street, Covent Garden, London WC2H 9HE

Library of Congress Control Number: 2026026786

British Library Cataloguing-in-Publication Data
A catalogue record for this book is available from the British Library.

GUIDE TO SEMICONDUCTOR ENGINEERING
Second Edition

ISBN 978-981-98-2603-2 (hardcover)
ISBN 978-981-98-2713-8 (paperback)
ISBN 978-981-98-2604-9 (ebook for institutions)
ISBN 978-981-98-2605-6 (ebook for individuals)

For any available supplementary material, please visit
https://www.worldscientific.com/worldscibooks/10.1142/14667#t=suppl

Desk Editors: Soundararajan Raghuraman/Steven Patt

Typeset by Stallion Press
Email: enquiries@stallionpress.com

To Ewa, Pawel and Julia

Preface

The shortage of adequately trained technical personnel to fill positions in the growing number of semiconductor manufacturing facilities around the world is a problem that can be addressed in two ways. First, engineers and technicians can be trained to meet the needs of specific manufacturing facilities and processes. Second, students in academic institutions can get an education in the broadly understood area of semiconductor engineering which is the foundation upon which the semiconductor industry is based.

This second edition of the *Guide to Semiconductor Engineering* was configured and written with both these needs in mind. Based on the author's 50 years' experience in semiconductor engineering research and teaching, the topics covered in this book, and the way in which they are presented, are specifically compatible with its goal of covering fundamentals members of semiconductor community should be familiar with. This knowledge of the basics of semiconductor engineering is the foundation upon which work-related specialization in any area of semiconductor technology can be built.

In an easy-to-follow fashion, which does not require either extensive engineering or scientific background, this book explains what the properties of semiconductor materials are distinguishing them from other solids, introduces key semiconductor devices, and discusses major elements of semiconductor process technology. It then considers the fabrication processes used in semiconductor device engineering and the ways to control them by monitoring properties of processed materials. To assist in educational needs of this second edition, a list of review questions is included at the end of each chapter.

It is with pleasure and gratitude that the author acknowledges the teachers, colleagues, and graduate students with whom he has had the privilege of interacting throughout the years. Among many of them, special thanks go to Professor Andrzej Jakubowski of the Warsaw University of Technology in Poland, Professor Junichi Nishizawa of Tohoku University in Japan, and to Penn State's mentors and friends Professors Joseph Stach and Richard Tressler. Other than that, this project would never come to fruition if it were not for the patience, understanding, and support of my wife Ewa.

Jerzy Ruzyllo
Penn State University, USA

About the Author

Jerzy Ruzyllo is a Distinguished Professor Emeritus in the School of Electrical Engineering and Computer Science at the Pennsylvania State University. He joined Penn State in 1984 after completing his education, obtaining PhD and DSc degrees in 1977 and 1984 respectively, and serving on the faculty of Warsaw University of Technology in Poland, where he currently holds the title of Honorary Professor. Throughout his career, which included two years' work at Tohoku University in Japan and half a year at Imec in Belgium, Dr. Ruzyllo was actively involved in research and teaching of semiconductor engineering-related topics.

Dr. Ruzyllo is a Life Fellow of IEEE and a Fellow of the Electrochemical Society. He authored the books titled *Semiconductor Glossary* in 2017, *Guide to Semiconductor Engineering* (1st Ed.) in 2020 and *Guide to Characteristics and Characterization of Semiconductor Surfaces* in 2025.

Contents

Chapter 1

Properties of Semiconductor Materials

1.1 Introduction

The goal of this chapter is to consider in qualitative terms fundamental properties of solids known as semiconductor materials. The discussion is not meant to be a structured overview of the physics of semiconductor materials, but rather a simplified review aimed at the identification of those properties of this class of materials which are essential to the understanding of the principles upon which functional semiconductor devices are built and work. More in-depth considerations of the physics of semiconductor materials are available in the literature (Pierret, 1996; Wolfe *et al.*, 1989; Grahn, 1991; Grundman, 2006).

The review starts with an identification of semiconductor materials among other solid-state materials based on their electrical conductivity. Then, the effect of the atomic structure of solids on their electrical conductivity, which is a basis upon which semiconductors are distinguished among other solids, is explained in an easy-to-follow fashion. Furthermore, the energy band structure of solids is considered and an important concept of the forbidden gap, also known as a bandgap, or energy gap, is introduced. The following section is devoted to electrons and holes acting as electric charge carriers in semiconductors and is followed by an overview of changes in semiconductor properties inflicted by external influences including electric field, magnetic field, light, and temperature.

The chapter is concluded with the discussion of the effect of nanoscale confinement of semiconductor geometry on its key characteristics.

1.2 Insulators, Conductors and Semiconductors

In materials engineering, the distinction of various classes of solids can be based on their mechanical, optical, thermal or electrical properties. The last option provides the basis upon which solids are considered in the discussion in this section in which their electrical conductivity is used as a reference.

In terms of electrical conductivity two extreme cases are represented by insulators, which are not conducting electricity (glass or rubber, for instance) and conductors featuring very high electrical conductivity such as metals. In both cases, under normal conditions, electrical conductivity of the material is little dependent on external influences such as electric field, light or temperature or on addition of alien elements to its structure.

Escaping this classification are materials referred to as semiconductors (referred to also as halfconductors). In contrast to insulators and conductors, semiconductors feature electrical conductivity, which is specific to any given material in this class. Consequently, various materials in this class of solids, either elemental or compound, can be distinguished based on their electrical conductivity (Fig. 1.1(a)). In addition, semiconductors are materials of which electrical conductivity can be altered in a controllable fashion by adding alien elements specific to any given semiconductor material in the process known as doping (Fig. 1.1(b)).

Key physical characteristics of semiconductors which determine electrical conductivity of the materials, and thus determine their usefulness in the construction of semiconductor electronic and photonic devices, are considered in more detail in the remainder of this chapter.

1.3 Electrical Conductivity of Solids

The extent to which a solid can conduct electricity is a characteristic defining its usefulness in the range of practical applications. In this section atomic-level features defining electrical conductivity of the solid, and at the same time are used as a criterion identifying a class of materials known as semiconductors, are considered.

1.3.1 *Interatomic Bonds and Electrical Conductivity*

Electrical conductivity σ, which is an inverse of resistivity ρ ($\rho = 1/\sigma$) and which is determined by the atomic structure of materials, represents

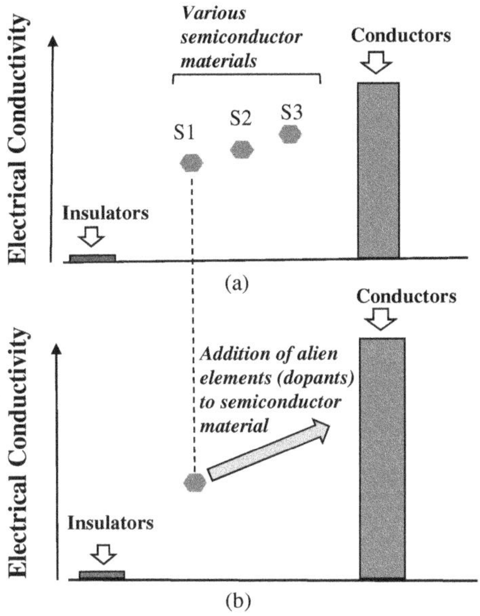

Fig. 1.1. Graphic representation of the features of semiconductor materials distinguishing them from insulators and conductors because in contrast to insulating and conducting materials (a) various semiconductor materials (S1, S2, S3, ...) feature distinctly different electrical conductivity and (b) electrical conductivity of semiconductor material (S1, for instance) can be increased in the controlled fashion by adding alien elements (doping process).

ability of material to conduct electricity. The atoms are comprised of positively charged protons and electrically neutral neutrons forming nucleus, which contains nearly all the mass of an atom, and negatively charged electrons in the number equal to the number of protons to assure electrical neutrality of an atom (Fig. 1.2). To promote the flow of electric current J across the solid, particles carrying electric charge must be set in motion by an electric field ε. Considering mass of an electron being negligible with respect to the mass of a nucleus, only electrons play a role of electric charge carriers in the solid.

As seen in Fig. 1.2 specific number of electrons is associated with atomic shells constituting an outside part of an atom around the nucleus. Not all electrons in the atoms can be set free and available for conduction. Electrons in the inner shells K and L in Fig. 1.2(a) are electrostatically strongly attached to the nucleus from which under normal conditions cannot be separated, and hence, cannot contribute to electrical conductivity.

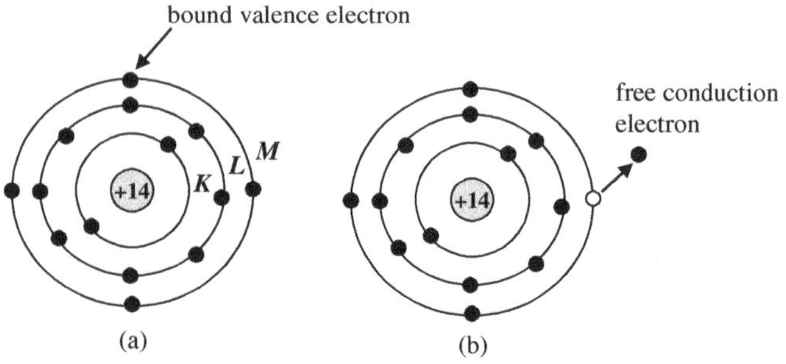

Fig. 1.2. (a) Schematic depiction of an atom using silicon with 4 electrons in the outermost shell *M* as an example, (b) valence electron acquires energy sufficient to overcome binding energy and to become a free conduction electron.

Only electrons in the outermost shell *M*, and known as valence electrons, can be separated from the nucleus upon receiving energy in excess of the energy binding them to the nucleus and become free conduction electrons (Fig. 1.2(b)).

What ensues from the above considerations is that the main factor controlling electrical conductivity σ of the solid is the number of electrons which are free, and hence, available for conduction in the presence of an electric field $\mathcal{E}$ enforcing electric current $J = \sigma \mathcal{E}$.

Availability of unbound from nucleus free electrons in the solid and the concentration (number per unit volume) of such electrons are the key factors defining its electrical conductivity. Good conductors of electricity such as metals feature essentially unlimited number of free electrons available to carry electric current. The number of electrons in metals cannot be reduced, and hence, these materials cannot be made less conductive. On the other end of the conductivity spectrum are insulators featuring lack of free electrons in their atomic structure. Here, due to the fundamental material characteristics, free electrons cannot be under normal conditions generated, and hence, such material can only be an insulator.

Semiconductors: In between two extreme cases defined above in terms of electrical conductivity there are solids referred to as semiconductors in which number of free electrons, and hence, their electrical conductivity can be altered and controlled within a broad range by introducing small amounts of properly selected alien elements (Fig. 1.1(b)). This characteristic is a fundamental property of semiconductor materials (Berger, 1997) distinguishing them from among other solids and making them uniquely suitable for the manufacture of the range of highly functional electronic and photonic devices which define the progress of our technical civilization since the mid-20th century. Adding to the versatility of semiconductor materials in practical applications is dependence of their electrical conductivity on illumination, temperature, as well as electric and magnetic fields all of which have little effect on electrical conductivity of metals and insulators.

Any consideration of the fundamental properties of materials needs to begin with the discussion of the nature of interatomic bonds involved. In contrast to gases and liquids, the solid-state materials under normal conditions in terms of temperature and pressure feature stability of shape. This stability of shape results from the strong electrostatic bonds (Coulomb forces) established between atoms closely spaced within the solid. The bond is formed when adjacent atoms are at the equilibrium distance at which repulsive and attractive forces acting upon the atoms are balanced and the force is zero.

The nature of bonds responsible for the cohesion of solids determines their fundamental properties including electrical conductivity. To illustrate this point, let us consider two different mechanisms of bond formation resulting in the materials featuring distinctly different electrical characteristics. The difference in bond formation mechanism results from the different number of valence electrons available to form a bond in various solids. As an example, we consider silicon and aluminum featuring four and three valence electrons respectively.

Covalent bond: Figure 1.3(a) shows two separated from each other silicon atoms schematically represented by positively charged atom cores and four valence electrons associated with the outer-most atomic shell. When brought into close contact, atoms will form a bond using two valence electrons, one from each atom participating in the bond as (Fig. 1.3(b)). A bond created this way is referred to as a covalent bond. To form covalently bonded lattice all four valence electrons of an atom are used to form bonds with four neighboring atoms leaving no free electrons. Measurable energy is required to break a covalent bond and to release an electron making it available for conduction (Fig. 1.3(c)). Materials featuring covalent bonds are not expected to be good conductors of electricity and as such are commonly referred to as semiconductors.

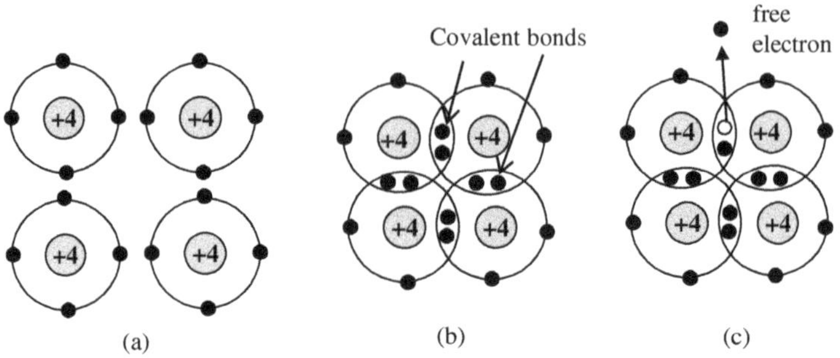

Fig. 1.3. Silicon atoms (a) separated, (b) brought together and forming covalent bond, (c) electron released from the covalent bond leaving behind positively charged hole which is free to move and to carry positive charge.

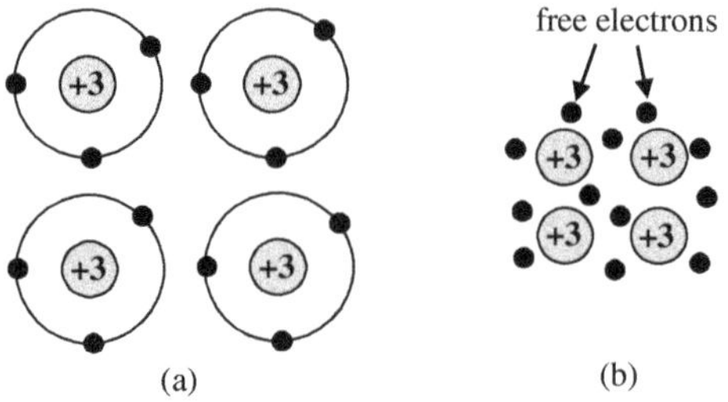

Fig. 1.4. Aluminum atoms (a) separated and (b) brought together and forming metallic bond.

The result of the electron being released from the covalent bond (Fig. 1.3(c)) is a left-behind hole which has the same charge as electron, but positive electric charge. Similarly to electron, the hole can move in semiconductor in the presence of electric field, and thus, can act as a carrier of the positive electric charge.

Metallic bond: Situation is different in the case of aluminum atoms featuring three valence electrons (Fig. 1.4(a)). Here, when atoms are brought into contact, valence electrons are taken away from each atom to form a community electron sea (Fig. 1.4(b)).

This electron sea as a whole interacts with the positive nuclei dispersed in it, and the resulting bond is referred to as a metallic bond. Since electrons in this case can move about upon receiving very little energy, the materials featuring metallic bonds are very good conductors of electricity such as metals.

Ionic bond: Only a limited number of elemental solids are purely covalent. In many binary (compounds) solids, covalent bonds are accompanied by the ionic bonds, which result from the attractive electrostatic forces between oppositely charged ions comprising the solid. Participation of ionic bonds usually further increases the energy required to release valence electrons and accounts for still lower electrical conductivity of these solids. In purely ionic solids, no electron conduction is possible, and the current is solely ionic in nature which means that for all practical reasons such solids are insulators.

1.3.2 *Energy Band Structure and Electrical Conductivity*

According to the rules of quantum mechanics the energy of electrons in a single atom is restricted to discrete levels. This implies that electrons by obeying the Pauli exclusion principle are allowed to occupy some energy levels in the atom and are not allowed to occupy others. In the periodic lattice of the solid featuring very closely spaced atoms, the discrete energy levels associated with single atoms are broadened into the bands, or ranges of energy allowable for electrons (Ferry, 2013). These bands are separated by the forbidden energy bands, or simply energy gaps in which no electron energy level is allowed. As can be intuitively expected, the energy band structure of solids, or in other words distribution of energy levels electrons can and cannot occupy, is a derivative of the nature of interatomic bonds which, as the discussion above indicated, determine energy state of electrons in the solid. Motion of electrons is restricted in the covalently bonded solids while in the solids featuring metallic bonds electrons are free to move around.

In the discussion of electrical conductivity of solids of interest is only the outermost section of the energy band structure consisting of valence bands and conduction bands depicted in Fig. 1.5 separated by the energy gap as shown in Fig. 1.5(a). The energy difference between edges of the conduction (E_c) and valence (E_v) bands respectively is a measure of the energy gap width E_g also referred to as a band gap width (Ruzyllo, 2016). The valence electrons have to acquire energy higher than the energy gap width to become conduction electrons, and hence, the energy gap width is

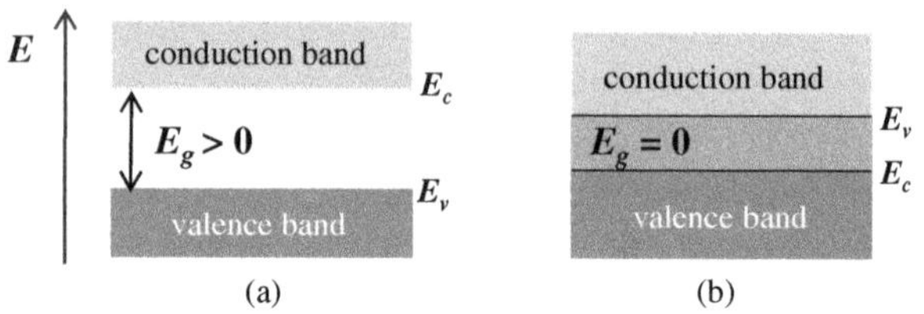

Fig. 1.5. Energy bands diagrams representing (a) semiconductors and insulators, and (b) metals.

a key factor determining electrical conductivity of solids. This, however, applies only to the solids in which the valence band is completely filled with electrons. In some solids, only the lower energy states within the valence bands are filled. In these solids, very little energy is sufficient to excite electrons from their ground state to the energy state where they become available for conduction. This means partial overlapping of the valence and conduction bands (Fig. 1.5(b)) and results in the lack of energy gap. In other words, essentially any electron in the valence band can contribute to the conduction. Such a situation reflects characteristics of the metallic bonds in solids (Fig. 1.5(b)) and is typical for metals.

Let us consider now solids which feature valence bands filled with electrons (Fig. 1.5(a)). Here, the valence electrons must overcome the energy gap to become conduction electrons. Otherwise, considering completely occupied valence bands, no "tied up" valence electron can carry electrical current. In consequence, the energy in excess of E_g, or in other words high enough to break interatomic bonds and release valence electrons, must be supplied to the solid to make it electrically conductive. Following this reasoning, the value of E_g can be used as a measure defining ability of various solids to carry electric current. Depending on the solid, the energy gap width may vary from zero eV for metals (Fig. 1.5(b)) to as high as 10 eV for insulators where electron volt (eV) is a unit of energy universally adopted in semiconductor terminology. In general, as there are exceptions to this rule, large energy gap materials (E_g >~5 eV) are insulators which do not allow electrical conduction at room temperature and which under normal conditions cannot be converted into more conductive solids. In contrast, solids featuring narrower energy gaps (E_g <~5 eV), and known as semiconductors, allow manipulation of their electrical conductivity over the several orders of magnitude and are conductive at room temperature.

Energy gap (bandgap) of semiconductors: The concept of energy gap, also referred to as a forbidden gap or a bandgap, as well as its characteristics play a pivotal role in determining how any given semiconductor material is used to fabricate electronic and photonic devices. The energy gap both in terms of its width E_g as well as alignment or misalignment of the maximum energy in the valence band and minimum energy in the conduction band with respect to each other define key characteristics of semiconductor material.

Fundamental concepts pertinent to the role of the band gap in defining characteristics of semiconductors are considered below.

Width of the energy gap E_g: Shown in Fig. 1.6(a), E_g is a material parameter which in many ways defines physical properties of semiconductor material. It determines its electrical conductivity, its tolerance to temperature, as well as its interactions with electromagnetic waves (light) in terms of both generation of light within semiconductor structure and absorption of light. Considering changes in those characteristics as the bandgap E_g increases, it is common to distinguish somewhat arbitrarily semiconductors featuring energy gap E_g > 2.5 eV as the wide-bandgap semiconductors.

Fermi level: An important reference needed in any consideration of the distribution of the energy levels in a solid that are available to electrons is a concept of Fermi level. It is defined as a level which can be occupied by the electron with a probability of 0.5 (Fig. 1.6(b)). Its position within the bandgap changes depending on the carrier concentration and temperature.

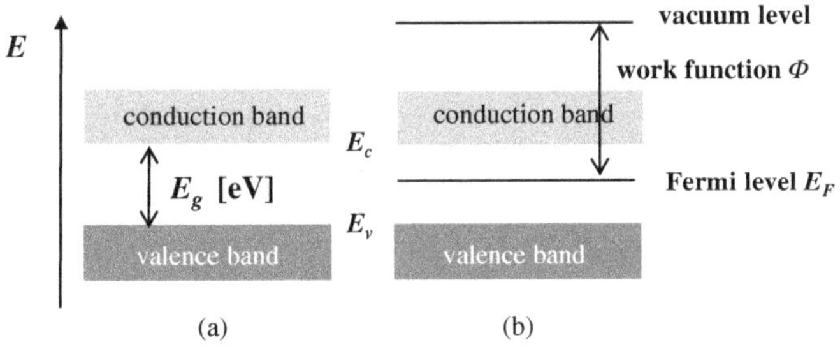

Fig. 1.6. Definition of (a) energy gap E_g and (b) work function Φ as key material parameters determining applications of any given semiconductor material.

Intrinsic semiconductor: In the case of a perfectly chemically pure and defect-free semiconductor material, referred to as an intrinsic semiconductor, the Fermi level is positioned exactly in the middle of the energy gap. Concentration of charge carriers in such a case is referred to as an intrinsic carrier concentration, n_i. Any departure from the intrinsic conditions makes the Fermi level move away from the middle of the bandgap either toward valence band or conduction band edges depending on the nature and extent of chemical and/or physical alteration of semiconductor crystal.

Work function: Energy needed to move an electron from the Fermi level to the vacuum level, which refers to the energy level outside of the atom, is called a work function (Fig. 1.6(b)). Work function is a material parameter which is an important reference when two solids featuring different chemical composition, and hence feature different distribution of the energy states, are brought to physical contact. The difference in work function between two solids in physical contact is responsible for the formation of the potential barrier in the contact area which controls the current between two materials in contact (see additional discussion in Chapter 3).

Direct and indirect bandgap: A simplified representation of the energy band structure of semiconductors shown in Fig. 1.6 does not allow demonstration of the features of the energy band structure affecting electron transitions between valence and conduction bands. To see how minimum energy in conduction band aligns with maximum energy in the valance band, distribution of the energy states needs to be considered as a function of crystal momentum, or *k*-vector (Fig. 1.7). In the case when maximum energy in the valence band and minimum in the conduction band occur at the same *k*-vector, then the bandgap is referred to as a direct bandgap (Fig. 1.7(a)). When the respective maximum and minimum energy do not coincide in the *k*-vector space then the energy gap is referred to as indirect bandgap (Fig. 1.7(b)). At the set chemical composition of semiconductor material direct or indirect nature of the energy gap is established and cannot be altered.

Bandgap engineering: For any given semiconductor material featuring set chemical composition and remaining at constant temperature, width of the energy gap E_g is predetermined and not subject to variations. By the

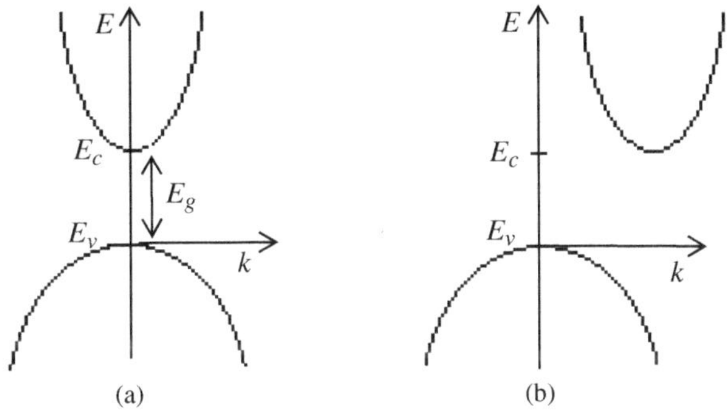

Fig. 1.7. (a) Direct bandgap and (b) indirect bandgap semiconductor materials.

same token width of the energy gap will change when chemical composition of the material is changed by, for instance, changing single-element semiconductor (elemental semiconductor) into two-element (binary) compound semiconductor material. Thus, if desired due to the device design requirements, width of the energy gap of semiconductor material can be altered by the modifications of its chemical composition. In the process known as bandgap engineering gradual changes of energy gap width can be accomplished in multilayer semiconductor structures where each layer features slightly modified chemical composition. In addition, there are cases where with proper engineering of semiconductor material's chemical composition the nature of the bandgap in semiconductor can be changed from direct to indirect and vice versa.

The width of the bandgap in semiconductors is independent of the volume of material as long as material geometry is not reduced to nanometer (atomic) scale dimensions. In the case of extreme geometrical confinement electrons start interacting with each other and their behavior in the solid can no longer be described by the laws of classical physics. In such situations, slight variations in the geometry of nano-confined materials results in the changes of their energy gap width. See Section 1.6 in this chapter for additional discussion regarding the effect of nanoscaling on fundamental characteristics of semiconductors.

1.4 Electric Charge Carriers

To allow electrical conductivity, free to move around carriers of an electric charge must be available in the body of the solid. Concentration of charge carriers (number of carriers per cm^3), the way they are generated, and the mechanism of their transport across the solid play the key role in defining electrical conductivity of any semiconductor material.

1.4.1 *Electrons and Holes as Carriers of Electric Charge*

As indicated earlier (Fig. 1.3(c)) there are two types of electric charge carriers in semiconductors: electrons carrying elementary negative charge q ($q = 1.602 \times 10^{-19}$ coulombs) and holes which carry the same charge, but positive. This contrasts with metals where there are no holes acting as charge carriers and electrons are solely responsible for carrying electric current. As discussed below, semiconductors in which concentration of electrons n exceeds concentration of holes p ($n \gg p$) are referred to as n-type semiconductors. In the same way semiconductors in which concentration of holes p exceeds concentration of electrons ($p \gg n$) are referred to as p-type semiconductors.

Besides different sign of the electric charge associated with electron and hole, the effective mass m^* of an electron is significantly smaller than the effective mass of the hole. This feature has a major impact on the electrons and holes charge carrying characteristics rendering the latter significantly less effective in this capacity.

Ability to control conductivity type of semiconductor material, n-type, or p-type, as well as concentration of charge carriers n and p is at the very core of semiconductor device manufacturing technology.

n-type and p-type semiconductors: In the case of an intrinsic semiconductor, concentrations of electrons n and holes p are by definition equal, $n = p$, and product $np = n_i^2$ where n_i depicts intrinsic carrier concentration. From the point of view of functional semiconductor devices, however, having a way to independently control concentration of electrons and holes is essential. In other words, there is a need for semiconductor materials in which either concentration of electrons is much higher than concentration of holes, $n \gg p$, and hence, electrical conductivity is driven by electrons, or semiconductor materials in which concentration of holes is much higher than concentration of electrons, $p \gg n$ and where electrical conductivity is driven by the holes. As mentioned earlier, in universally adopted terminology the former are known as n-type semiconductors and the latter as p-type semiconductors. In the case of n-type semiconductors, electrons are acting as majority carriers and holes as minority carriers. Accordingly, in p-type semiconductor holes are majority carriers and electrons act as minority carriers. Note that in equilibrium, regardless of whether semiconductor is n-type or p-type, product np remains equal to n_i^2.

Control of conductivity by doping: To make any given semiconductor n-type or p-type, properly selected alien elements need to be added to the host semiconductor material. The process is known as doping and alien elements added are referred to as dopants (sometimes term impurities is used instead). In contrast to the discussed earlier modifications of semiconductor material's chemical composition carried out for the purpose of bandgap engineering, doping requires only minute alterations of the chemical makeup of the host material. In fact, one dopant atom per million host atoms is enough to observe noticeable changes in semiconductor's conductivity type n, or p.

Selection of dopant atoms is based on the number of valence electrons featured by the host atoms. If, for example, the host atoms feature four valence electrons (Fig. 1.8), then the element with five valence electrons needs to be introduced into the lattice to add free electrons and make host material n-type. Dopant atoms are referred to in this case as donors and their concentration is denoted as N_D. In the course of the doping process a dopant atom substitutes for the host atom in the lattice then uses its four electrons to form covalent bond with adjacent host atoms leaving one free electron ready to act as a charge carrier (Fig. 1.8(a)). Under the equilibrium condition concentration of electrons in n-type semiconductor $n = N_D \gg p$ and electrons are acting as majority carriers.

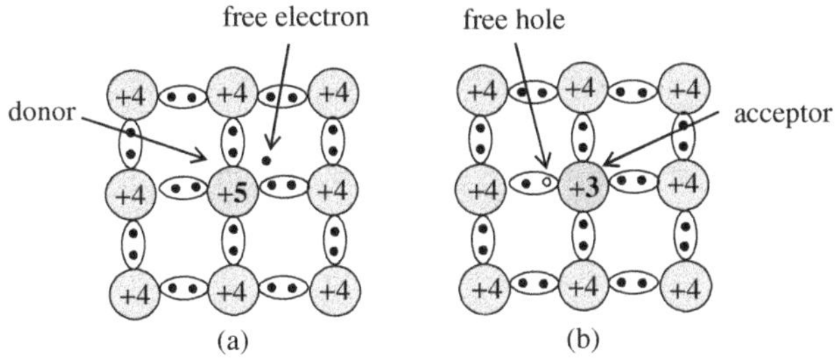

Fig. 1.8. (a) n-type doping (donor), (b) p-type doping (acceptor).

To make host material *p*-type rather than *n*-type, dopant atoms featuring three valence electrons are used. In this case, to form a covalent bond with adjacent atoms in the lattice dopant atom needs to accept an electron from the existing bonds leaving behind a hole which is free to move around and to contribute to electrical conductivity of the host material and making it *p*-type (Fig. 1.8(b)). Dopant atoms establishing a *p*-type semiconductor are referred to as acceptors and their concentration is denoted as N_A. Under the equilibrium condition concentration of holes in *p*-type semiconductor $p = N_A \gg n$ and holes are acting as majority carriers.

The electrical conductivity of *n*-type and *p*-type semiconductors is no longer solely dependent on their intrinsic properties, and accordingly, doped semiconductors are referred to as extrinsic semiconductors. The energy needed to make donor atom to "donate" one of its valence electrons, and hence, to acquire a positive charge and become an ion, is known as ionization energy. The same term is used in reference to the acceptor atoms which are negatively ionized by accepting one extra electron.

1.4.2 *Generation and Recombination Processes*

Generation and recombination processes are controlling availability of free charge carriers in semiconductors. Generation is the process of free charge carriers' formation in semiconductor materials resulting from the electron acquiring energy from outside of the atom, for instance, thermal energy or energy of light. The energy supplied to semiconductor needs to be sufficient to overcome energy gap and to allow electron's transition from the valence band where it cannot move, to the conduction band where it can move, and thus, contribute to the electrical conductivity of material. An outcome of the process is an empty state left in the valence band known as a hole (Fig. 1.9(a)). Hence, a band-to-band generation process results in a generation of electron–hole pairs instead of just electrons. To reiterate points made earlier, hole features an electric charge the same as electron, but positive and just like electron in the conduction band, hole can move around in the valence band acting as a carrier of positive charge in semiconductor materials.

The process of recombination is opposite to generation and results in annihilation of the electron–hole pair by electrons releasing energy equivalent to the energy gap E_g and transitioning from the conduction to the valence band where they recombine with holes (Fig. 1.9(b)). The energy resulting from recombination can be released either in the form of light,

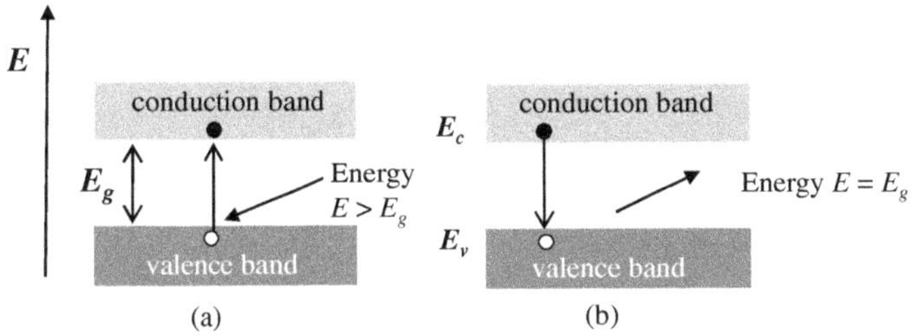

Fig. 1.9. (a) Generation of the electron–hole pair as a result of band-to-band generation and (b) annihilation of the electron–hole pair through the band-to-band recombination.

or in the form of the heat wave dissipated into the lattice of semiconductor depending on whether energy gap of semiconductor material is direct or indirect.

Minority carrier lifetime: The time between charge carrier generation and recombination is called carrier lifetime, τ, and is denoted as τ_n and τ_p for the lifetime of electrons and holes respectively. The minority carrier lifetime is strongly dependent on the density of recombination centers in semiconductor material which most commonly are related to structural defects in semiconductor crystal. Consequently, measurements of the minority carrier lifetime can be used to estimate density of such defects in semiconductor crystals. In highly defective crystals the minority carrier lifetime can be as short as a fraction of a microsecond, while in high-quality crystals can be in the high millisecond range.

Overall, generation and recombination processes play an important role in various facets of the operation of essentially any semiconductor device as they influence the number of free charge carriers available for conduction.

1.4.3 *Charge Carriers in Motion*

To carry electric current, charge carriers need to be set in motion. Factors affecting transport of charge carriers in semiconductors are considered in the following discussion.

Mobility of charge carriers: Free electrons and holes carrying an electric charge and moving in semiconductor material are subject to severe scattering resulting from collisions and electrostatic interactions with host and dopant atoms in the lattice, as well as with structural defects. All these interactions come down to the material specific alterations of the charge carrier movement in semiconductors. Quantitatively, these alterations are collectively reflected in the changes of the value of the material parameter known as mobility, μ (unit cm^2/Vs) of charge carriers. Mobility of charge carriers has a major effect on electrical conductivity of any semiconductor material (Lundstrom, 2000). As such, it predetermines material's usefulness in specific device applications

Due to the significant impact of charge carrier mobility on the design and performance of semiconductor devices, correlation between material characteristics and carrier mobility needs to be well understood and controlled.

The impact of selected semiconductor material characteristics on charge carrier mobility is graphically summarized in Fig. 1.10. First off, due to the lower effective mass of electrons than holes, electron mobility μ_n is in general higher than the mobility of holes μ_p. As a result, it is a common practice to configure semiconductor devices such that faster moving electrons rather than slower holes are responsible for conduction in certain parts of the device. The effective mass of electrons and holes is different in different semiconductors, and hence, mobilities of electrons and holes are different in different semiconductors and as such are material specific parameters. In this context, semiconductors featuring high electron mobility are referred to as high electron mobility materials.

Also, increased dopant concentration lowers mobility because of increased scattering on the dopant atoms in the lattice. In addition, mobility decreases with temperature because of the increased vibrations of the atoms in the lattice which increases scattering of moving charge carriers.

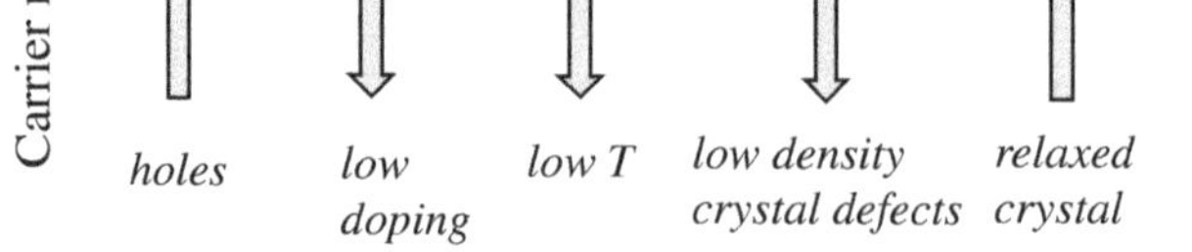

Fig. 1.10. Changes of charge carriers' mobility in semiconductors depending on material characteristics.

Furthermore, charge carrier mobility depends on the crystallographic structure of the material (Fischetti, 2016). It is significantly higher in the structurally ordered semiconductor materials than in structurally disordered semiconductors featuring the same chemical composition (see Chapter 2 for more detailed discussion of crystallographic structure of semiconductors). Structural defects in the crystal cause enhanced scattering of the moving charge carriers thereby lowering their mobility. For that reason, charge carriers moving in the structurally disturbed near-surface region of crystalline semiconductor feature lower mobility than in the structurally undisturbed bulk of the same material (Ruzyllo, 2025). On the other hand, in crystalline material in which crystal lattice is strained due to the stress induced by lattice deformation, an effective mass of electrons is lower than in the relaxed lattice, and hence, electron mobility is higher.

Considered earlier equation defining current density in solids as $J = \sigma\mathcal{E}$ did show current being proportional to electrical conductivity of a solid σ and electric field $\mathcal{E}$. We understand now that in the case of semiconductors, electrical conductivity, or in other words, the ability to move elementary electrical charge q around, depends on the concentration of electrons n and their mobility μ_n, in n-type semiconductor ($\sigma = q\mu_n n$), and concentration of holes p and their mobility μ_p, in p-type semiconductors ($\sigma = q\ \mu_p p$).

It needs to be noted at this point that in everyday semiconductor terminology the resistivity ρ, being an inverse of conductivity ($\rho = 1/\sigma$), rather than conductivity σ is often used as a parameter defining correlation between doping and electrical conductivity of semiconductors.

Charge carrier transport mechanisms: Under the normal conditions in terms of geometrical confinement of semiconductor material and temperature, there are two mechanisms that can initiate electric current resulting from the net flow of electrons and/or holes in semiconductor.

The first is drift (drift current) which is a movement of charge carriers driven by the electric field $\mathcal{E}$. Drift is an electric current producing mechanism not only in semiconductor materials, but also in metal conductors wherever in our daily life electricity is used.

The second mechanism, which does not require electric field, and which is specific to semiconductor materials, is a diffusion (diffusion current). In this case the flow of carriers is driven by concentration gradient, or in other words non-uniform distribution of charge carriers in semiconductor material. The current in this case flows in the direction of the lower concentration region until uniform distribution of carriers is reached.

If the non-uniform distribution of charge carriers is maintained, for instance, by continued injection of charge carriers into one end of semiconductor region, then the flow of diffusion current will continue for as long as the mechanism causing non-uniform distribution of charge carriers will remain active.

Whether drift current or diffusion current controls operation of any given semiconductor device depends on the principles of operation of such device and its design. In general, drift current underlies operation of the unipolar devices while diffusion current is a device controlling current flow mechanism in bipolar devices. Discussion in Chapter 3 of this volume considers these two classes of semiconductor devices in more detail.

Velocity saturation: Velocity of charge carriers moving in semiconductors under the influence of electric field (drift velocity) increases with the increasing electric field and saturates at certain maximum values. Saturation occurs because of the excessive scattering of charge carriers drifting in semiconductor lattice with very high velocity. Saturation velocity and electric field at which it is reached are material parameters which are different for different semiconductor materials due to differences in different semiconductor materials spatial distribution of atoms in the crystal lattice. Values of the saturation velocity and electric field at which it saturates are good predictors of the ability of semiconductor material to operate under the high electric field conditions.

1.5 Semiconductor Materials and External Influences

In contrast to other solids, certain key physical characteristics of semiconductor materials are altered when material is exposed to externally applied electric and/or magnetic field as well as to light or temperature. This sensitivity to external conditions underlies the operation of several important semiconductor devices.

1.5.1 *Semiconductor Materials and Electric and Magnetic Fields*

When electric field or magnetic field are present in semiconductor, its conductivity is affected in the well-understood and predictable fashion.

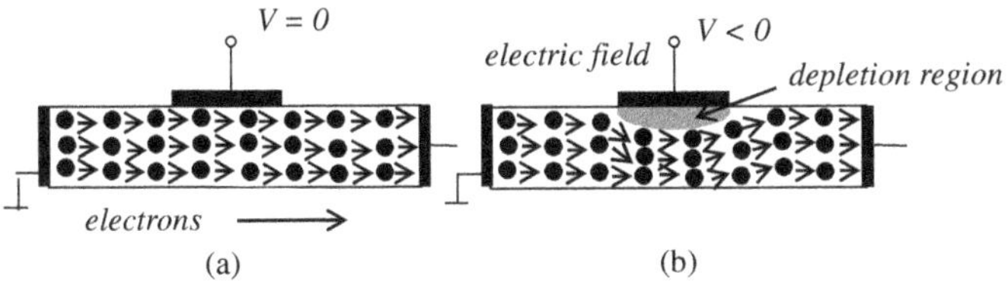

Fig. 1.11. Schematic illustration of the field effect. (a) electrons drift current in *n*-type semiconductor sample is (b) altered by the negative potential applied to the surface of the sample.

Field-effect: Control of electrical conduction of semiconductor in the geometrically confined regions, typically near its surface, can be accomplished via the field-effect. A field-effect is initiated by the potential applied to semiconductor surface in the way causing repulsion or attraction of the free charge carriers in the direction normal to the surface (Fig. 1.11). In this way conductivity of the near surface region underneath the surface potential establishing electrode can be controlled by the voltage V as shown in Fig. 1.11.

The field-effect is broadly exploited in the operation of semiconductor devices known as field-effect devices which are discussed in Chapter 3 of this volume.

Hall effect: Magnetic field alters characteristics of semiconductors by interacting with charge carriers in motion. A prime manifestation of these interactions is a Hall effect that causes generation of potential difference (the Hall voltage) across semiconductor sample in the x direction which is perpendicular to the current flow in the y direction in the presence of magnetic field applied in the z direction (Grahn, 1991; Grundman, 2006).

The Hall effect finds its applications in semiconductor sensors (see Chapter 3) and in semiconductor material characterization, for instance, in the measurements of charge carrier mobility (see Chapter 6).

1.5.2 *Semiconductor Materials and Light*

Interactions involving semiconductors and light encompass two key functions that can be performed by semiconductor materials structured

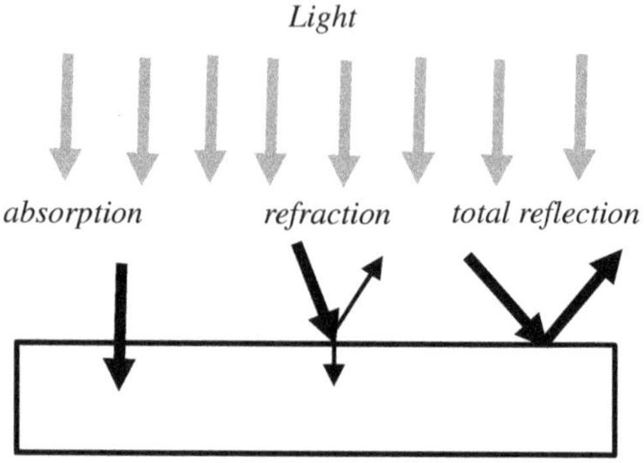

Fig. 1.12. Light from the outside source illuminating surface of the solid may be subjected to absorption, refraction, or total reflection.

into functional devices: conversion of the absorbed light into electric current and conversion of the electric current into light emitted by semiconductor device. In both these cases physical phenomena of absorption, reflection and refraction define interactions of light with solids as shown in Fig. 1.12.

Among them absorption allows penetration of energy carried by light into the exposed semiconductor material, reflection is a well understood concept while the phenomenon of refraction refers to the change in the direction of light passing through the boundary between two media featuring different optical properties (Fig. 1.12). For instance, the beam of light propagating in the air at a certain angle with respect to the illuminated surface of the solid will change its direction upon penetrating the solid (as opposed to being reflected).

A refractive index n is a material parameter of any given solid defining its interactions with light. An interplay between light reflection and refraction defines some key characteristics of semiconductors' interactions with light.

Conversion of light into electric current, e.g. solar cell: Energy carried by electromagnetic radiation in the form of light is $E = h\nu$ where h is the Planck constant and ν is a frequency of electromagnetic wave. After converting hv into wavelength λ related relationship, correlation between the energy of light and its wavelength is expressed as E (eV) = 1.24/λ (μm).

The light illuminating semiconductor material carries energy E which is absorbed in semiconductor when $E > E_g$, i.e. when energy of light is larger than the energy gap of illuminated semiconductor. The light passes through semiconductors unabsorbed, i.e. semiconductor material is transparent, when $E < E_g$. In the former case, absorbed light carries enough energy to initiate discussed earlier band-to-band generation of electrons and holes and to contribute in this way to the increase of the current flowing in the semiconductor through the effect of photoconductivity. As an example, semiconductor featuring bandgap $E_g = 1.1$ eV is absorbing UV light featuring wavelength $\lambda = 0.459$ μm because it carries energy $E = 2.7$ eV > 1.1 eV. On the other hand, the same semiconductor material is transparent to the infrared light featuring wavelength $\lambda = 1.77$ μm and carrying energy $E = 0.7$ eV < 1.1 eV.

The effect of photoconductivity occurs in semiconductors only and is a foundation of the light-to-electricity conversion using semiconductor devices including solar cells (see discussion in Chapter 3).

Conversion of electric current into light, e.g. light emitting diode (LED): A current flown into semiconductor as a result of applied voltage will bring along an increased concentration of free charge carriers which eventually will be subjected to recombination as discussed earlier in this section. Any recombination event, i.e. transition of electron from the higher energy level in conduction band to the lower energy level in the valence band is accompanied by the release of energy (Fig. 1.8(b)). Depending on semiconductor, the energy can be released in the form of (i) "packets", or quanta of electromagnetic radiation referred to as photons and forming a beam of visible or invisible light or (ii) in the form of the "packets" or quanta of energy of lattice vibrational wave referred to as phonons. For the former to occur, semiconductor must feature a direct band gap in which case band-to-band recombination of electrons with holes occurs without any momentum transfer and the energy released results primarily in the emission of photon with energy $E_{\text{photon}} = E_g$ (Fig. 1.13(a)). In the case of indirect gap semiconductor only small portion of energy is released as the energy of photon E_{photon} because recombining electrons must pass through an intermediate state and transfer momentum to the crystal lattice in the form of the energy of the phonon E_{phonon} (Fig. 1.13(b)).

The wavelength of light emitted by the direct bandgap semiconductor depends on its energy gap width E_g. With available semiconductors it may vary from the long-wavelength infrared (IR) to the short-wavelength ultraviolet (UV) covering the entire spectrum of visible light in between.

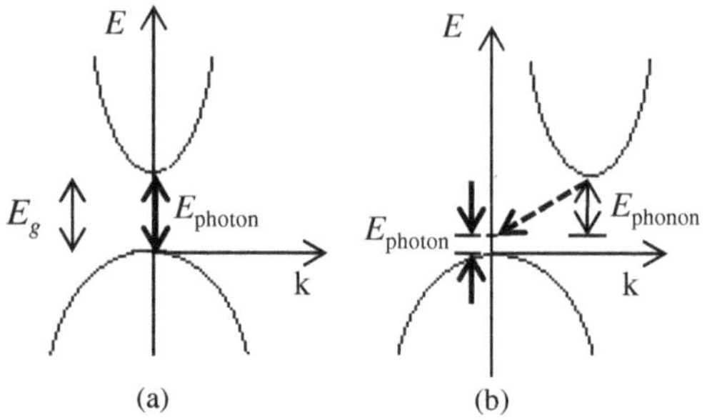

Fig. 1.13. Band-to-band recombination processes in (a) direct bandgap semiconductor resulting in the emission of photon (radiative recombination $E_g = E_{photon}$) and (b) indirect bandgap semiconductor where energy is released predominantly in the form of the phonon (nonradiative recombination, $E_g = E_{photon} + E_{phonon}$).

The wavelength of light λ emitted through band-to-band recombination in the direct band gap semiconductor can be calculated from the introduced earlier relationship $\lambda\ (\mu m) = 1.24/E_g\ (eV)$.

As an example, direct band gap semiconductor featuring bandgap $E_g = 2.7$ eV will emit radiation with a wavelength $\lambda = 0.459\ \mu m$ corresponding to the visible light. The direct band gap semiconductor featuring bandgap $E_g = 0.7$ eV will emit light with a wavelength $\lambda = 1.77\ \mu m$ falling into invisible infrared part of the electromagnetic spectrum.

1.5.3 *Semiconductor Materials and Temperature*

As it should transpire from the above considerations, temperature plays an important role in essentially all facets of physical characteristics of semiconductor materials. The impact of temperature can be qualitatively summarized by considering changes of electrical conductivity of the *n*-type semiconductor $\sigma = q\mu_n n$ resulting from the changes of electron concentration n and electron mobility μ_n as a function of temperature depicted in Fig. 1.14.

At the temperature of absolute zero, all electrons remain in the valence band, no free electrons are available in conduction band and semiconductor material is essentially non-conductive. As the temperature increases the process of free electrons generation through dopant atoms ionization rapidly intensifies and continues until at the thermal energy reaches ionization energy corresponding to temperature T_I at which point

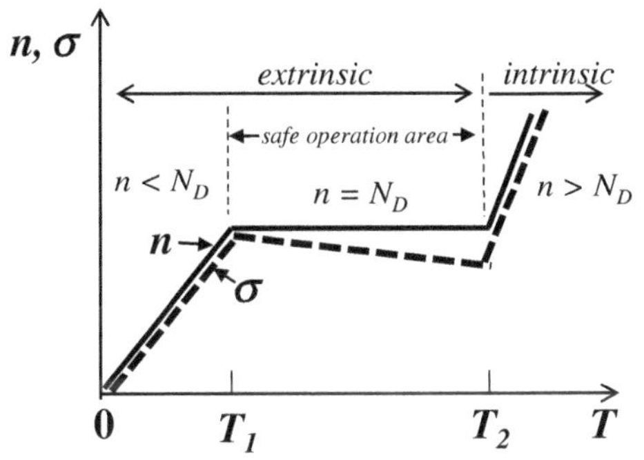

Fig. 1.14. Schematic qualitative representation of the changes of electron concentration n and conductivity σ of n-type semiconductor with temperature.

all dopant atoms are ionized, and hence, cannot supply any additional free electrons. In $T < T_1$ temperature regime, conductivity σ is dominated by the rapid increase of electron concentration n which overshadows the effect of decreasing electron mobility μ_n with temperature. As a result, in this temperature range changes of conductivity σ follow changes of electron concentration with temperature (Fig. 1.14). Once all dopant atoms are ionized at the temperature T_1, further increase of electron concentration is negligible until thermal energy becomes sufficient to initiate the process of electron–hole pair band-to-band generation (Fig. 1.9(a)) at temperature T_2 and above. In the interim ($T_1 < T < T_2$), conductivity of semiconductor material decreases as the mobility of electrons decreases with temperature while electron concentration n remains unchanged.

Once temperature reaches temperature T_2 corresponding to thermal energy higher than energy gap E_g, the process of electron–hole pair generation takes over control of electrical conductivity which rapidly increases as a result. At $T > T_2$ semiconductor material loses its extrinsic characteristics and assumes characteristics of an intrinsic semiconductor.

In addition to qualitatively demonstrating changes of electrical conductivity of semiconductors with temperature, Fig. 1.14 can also be used to define safe operation area of semiconductor materials in terms of temperature. Below temperature T_1 extrinsic semiconductors do not have established conductivity type because dopant ionization is not completed, i.e. they are not fully developed either n-type or p-type. Above temperature T_2 on the other hand, extrinsic characteristics are no longer sustained because the rapid increase of electron–hole generation renders semiconductor intrinsic, which means that the original either *n*-type or *p*-type conductivity is no longer in place. Based on these considerations, the temperature range assuring safe operation of semiconductor devices is between temperatures T_1 and T_2. The temperature T_2 increases as the bandgap width increases, and thus, wide-bandgap semiconductors are more resistant to temperature than their narrower bandgap counterparts.

What needs to be stressed is that the increased temperature of semiconductor material may not only be a result of external heating, but may also be a result of the inherent operational characteristics of some semiconductor devices. For instance, advanced semiconductor integrated circuit performing complex computational functions may potentially generate heat in excess of the temperature T_2 causing destruction of the circuit. In this context thermal conductivity of semiconductor, which defines its ability to dissipate heat, is coming to play. In general, thermal interactions involving semiconductor devices are at the core of the important part of semiconductor device engineering known as heat management.

1.6 Properties of Semiconductor Materials in Nanoscale

Unlike light and temperature below critical level (Fig. 1.14), which induce transient changes in semiconductor physical properties, reduction of the size of semiconductor material to the atomic scale dimensions, brings about permanent changes in its physical properties drastically affecting charge transport mechanism as well as light absorption and emission characteristics. For instance, the width of the bandgap of semiconductor materials is independent of the volume of material as long as material geometry is not reduced to the nanometer (atomic) scale dimensions. In the case of an extreme geometrical confinement, electrons start interacting with each other and their behavior in the solid can no longer be described by the laws of classical physics. In such situations, slight variations in the geometry of nano-confined semiconductor material results in changes in the energy gap width.

In order to attribute physical meaning to the term atomic scale, or nanoscale, let's remind ourselves that the diameter of an atom depends on the number of electrons it encompasses and varies from element to element from roughly 0.1 to 0.5 nanometer where nanometer, nm, is

1 billionth of the meter (1 nm = 10^{-9} m). The diameter of silicon atom shown in Fig. 1.2 is 0.22 nm. As a reference, the average size of bacteria is on the order of 1,000 nm, red blood cells about 6,000–8,000 nm. The only species in the bio-world sized in the low nanometer regime are viruses at about 20 nm.

The term *nanoscale* is being used throughout this volume in reference to materials, materials systems, or parts of the devices which along at least one out of the three dimensions, translated into the length, width, and thickness of the piece of material, are reduced to below some 10 nm. While recognizing the highly arbitrary nature of this numerical assumption it is believed that by assigning a specific number to the term *nanoscale* some important aspects of state-of-the-art semiconductor engineering will be easier to follow.

The point being made here is that the fundamental properties of a solid confined to ultra-small (nanoscale) geometries are different than the properties of the exact same solid in the bulk form, or in other words, in the form in which its physical properties are the same in all three dimensions and independent of the size of the sample. Figure 1.15(a) shows a crystalline three-dimensional (3D) bulk material, featuring well defined physical characteristics governed by the laws of classical physics. It is enough, however, that its one dimension is reduced to close to zero along the *z* axis, for instance, effectively forming a sheet few atoms thick (Fig. 1.15(b)), also referred to as nanosheet, that its properties change significantly. This is because two-dimensionality, or 2D-confinement, alters distribution of energy levels which can be occupied by electrons in the atom, and hence, changes fundamental physical properties which now are described by the laws of quantum physics. The behavior of electrons

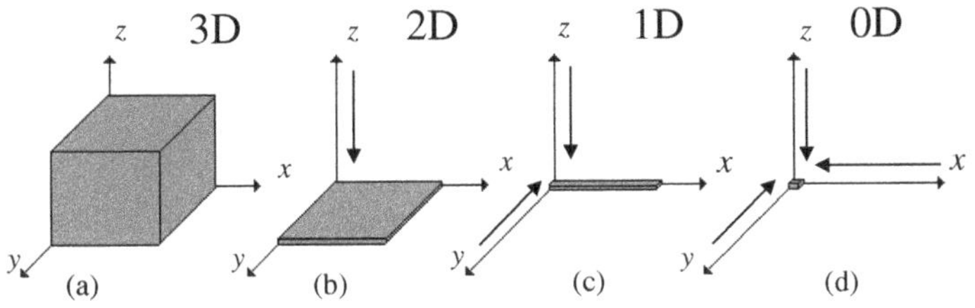

Fig. 1.15. Physical properties of semiconductor material as its geometry is changing from (a) not geometrically confined 3D sample to (b) 2D material system, (c) 1D material system, and (d) zero-dimensional (0D) material system.

in x and y directions is now very different than it was in exactly the same material in x and y directions in the bulk form (Fig. 1.15(a)). As it will be discussed in the remaining Chapters of this book, there are instances in which 2D-confined material systems play performance defining role in semiconductor devices.

If the same exercise as above is continued and the 2D geometry is reduced along the y axis (Fig. 1.15(c)), the resulting one-dimensional (1D) configuration of the same solid as in the cases (a) and (b) in Fig. 1.15 will feature even further modified physical characteristics. Nanoscale material systems referred to as nanowires and nanotubes are examples of the 1D material systems.

The modifications of the physical properties of a solid by manipulating its geometrical dimensions can be continued toward formation of the 0D material by reducing to close to zero the x axis (Fig. 1.15(d)). The resulting structures known as quantum dots, or nanodots in short, feature remarkable physical properties in many aspects drastically different from those displayed by the same material in 3D, 2D, or 1D configuration. For instance, in the case of semiconductors, nanodots show dependence of the width of the energy gap on the size of the dot allowing tunability of basic properties of the material by changing the size of the dot by fractions of the nanometer (Sabeeh *et al.*, 2019).

The ability to form nano-confined structures shown in Fig. 1.15 using semiconductor materials not only allows innovative approaches to the design of the existing devices but also opens new areas of application of semiconductor technology in general. The discussion in this *Guide* will emphasize this point further.

There are several types of nanoscale semiconductor materials with groundbreaking uses in the manufacture of transistors, flexible and/or transparent electronics and photonics, bio-gas sensors, and many other types of devices (see Chapter 4). The main reason for the interest in nanoscale materials is, as discussed in Section 1.5, related to physical phenomena such as tunneling or ballistic transport which are observed only when the size of semiconductor sample is reduced to the atomic-scale dimensions. The nanoscale confinement also results in the widening of the bandgap of semiconductors. Exploitation of these phenomena in semiconductor devices is expanding semiconductor engineering into previously uncharted territories.

Review Questions

1.1. Can electrical conductivity of semiconductor material be increased in the controlled fashion? If so, how can it be done? Can the same be done in the case of conductors, metals, for instance?
1.2. What feature of a solid material determines its electrical conductivity, and thus, decides whether it is an insulator, conductor or semiconductor?
1.3. Explain the concept of energy gap in semiconductor materials. How does it differ between insulators, conductors and semiconductors? What do direct and indirect energy gaps (bandgaps) mean?
1.4. In contrast to conductors (e.g. metals), semiconductor materials feature negative and positive charge carriers. What are the charge carriers motion mechanisms in semiconductor materials?
1.5. Generation-recombination processes affect concentration of electric charge carriers in semiconductor materials. Consider recombination processes in semiconductor materials featuring direct and indirect bandgap.
1.6. Describe charge carriers transport mechanisms in semiconductor materials.
1.7. Explain mechanism according to which electric field may alter electrical conductivity of semiconductor material in its near-surface region. Can the same be accomplished in insulators and conductors?
1.8. What needs to be the energy of light illuminating semiconductor material to increase its conductivity?
1.9. Define in general terms the temperature range in which semiconductor device can operate. Does this range depend on semiconductor material used to fabricate the device?
1.10. What happens when thickness of a solid (for instance, single-crystal silicon or carbon in the form of graphite) is reduced from the thickness of 3D sample to the thickness of one atom?

Chapter 2

Materials in Semiconductor Device Technology

2.1 Introduction

Following an overview of the properties of the solids known as semiconductors in the previous chapter, the goal of this chapter is to introduce semiconductor materials (Ruzyllo, 2016) which are used to manufacture devices performing variety of electronic and photonic functions. The discussion in this chapter distinguishes between inorganic and organic semiconductors and in the case of the former, considers elemental and compound materials displaying semiconductor properties.

Beside chemical composition, other materials related aspects of semiconductor device engineering are considered in the follow-up discussion. This includes topics related to the crystallographic structure of semiconductor materials both bulk and thin film and is followed by the review of the type of substrates upon which semiconductor devices are built.

Furthermore, the discussion in this chapter takes into consideration the fact that insulators and metals play an essential role in the engineering of functional devices based on physical properties of semiconductors, and thus includes a brief overview of insulators and metals used in semiconductor device technology.

2.2 Crystal Structure of Solids

The discussion of the crystal structure of solids is concerned with the way atoms comprising a solid are spatially distributed. In the case of semiconductors this issue is of fundamental importance as the way in which atoms in any given material are configured with respect to each other is among key factors defining electrical conductivity of semiconductors, and hence, defining the way any given semiconductor material is used to construct functional semiconductor devices.

An underlying consideration in this regard is the extent to which spatial distribution of atoms in the solid is ordered, and what is the geometrical nature of the ensuing crystallographic order. Two distinct cases are represented by (i) crystals featuring a long-range periodic order and (ii) non-crystalline materials, commonly referred to as amorphous materials in which, in contrast to crystals, atomic arrangement exhibits no periodicity or long-range order. Among crystals, single-crystal and poly-crystalline materials (in some situations also referred to as multicrystalline materials) are distinguished. In the former case, a periodic long-range order is maintained throughout the entire piece of material (Fig. 2.1(a)) while in the latter case such order is maintained only within the limited in volume grains which are randomly connected to form a solid (Fig. 2.1(b)). An amorphous, non-crystalline material does not feature a long-range order at all (Fig. 2.1(c)).

In semiconductor engineering the lead role is played by the single-crystal materials (Fornari, 2018) while polycrystalline/multicrystalline and non-crystalline amorphous materials are used in a range of specialized applications in support lower cost construction and fabrication of semiconductor devices and circuits.

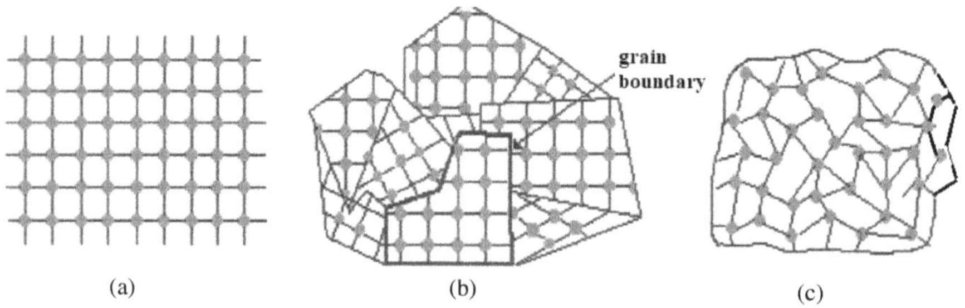

Fig. 2.1. 2D representations of crystallographic structure of solids, (a) single-crystal material, (b) polycrystalline/multicrystalline material, and (c) non-crystalline, amorphous, material.

2.2.1 *Crystal Lattice*

In the follow-up discussion the term crystal lattice is used in reference to the repeated 3D arrangement of atoms in the crystal. Any crystal lattice is comprised of the elemental cells reproduced throughout the material. Elemental cells may appear in a variety of forms which fall into seven basic classes. Several key semiconductor materials used to fabricate semiconductor devices such as silicon (Si) and gallium arsenide (GaAs) crystals, belong to the cubic class of crystals, and more specifically, they represent two different variations of the same face-centered cubic (FCC) cell (Fig. 2.2(a)). Due to the different nature of atomic bonds in silicon and GaAs the former adopts a diamond crystal lattice while the latter crystallizes in the zinc blende crystal lattice version of the face-centered cubic cell. In addition to the cubic class of crystals, some key compound semiconductor materials such as gallium nitride GaN, for instance, crystallize in the hexagonal crystal configuration represented in Fig. 2.2(b) by wurtzite crystal.

The physical dimension of the unit cell is defined by the lattice constants. In the case of cubic crystals, single lattice constant a (Fig. 2.2(a)) which in the case of silicon is a = 0.543 nm and for gallium arsenide a = 0.564 nm determines the lattice. In the case of hexagonal wurtzite crystal structure, two constants a and c are needed to define physical dimensions of the cell (Fig. 2.2(b)). As the discussion in Section 2.6 will demonstrate, lattice constant is a key characteristic of any given single-crystal material determining its structural compatibility with other single-crystal materials.

By connecting points on the main crystallographic axes of the single-crystal material, various crystallographic planes within the crystal can be

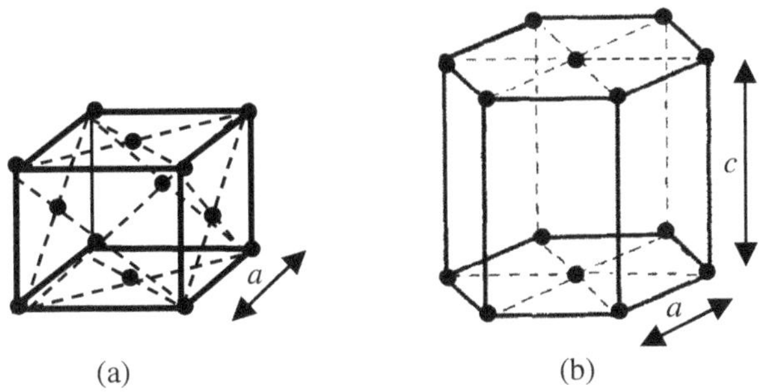

Fig. 2.2. Schematic representation of (a) face-centered cubic cell and (b) hexagonal cell.

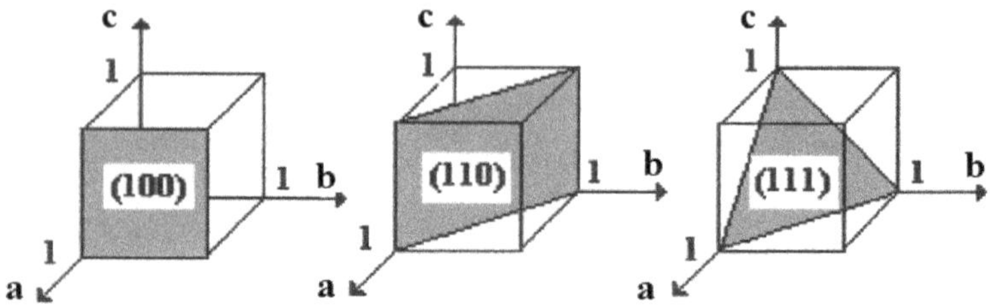

Fig. 2.3. Illustrations showing use of Miller indices to define crystallographic planes in the cubic cell.

identified. Coordinates determining orientation of the specific crystallographic plane in the crystal are known as Miller indices. Miller indices involve three digits, either 1 or 0, e.g. (111) or (100), which define how the plane intersects the main crystallographic axes of the crystal. Using cubic cell as an example, Fig. 2.3 shows how Miller indices define key crystallographic planes in this class of crystals.

2.2.2 *Structural Defects*

Considering the complexity of the crystal structure of semiconductors, an assumption that they consist of the perfectly periodic 3D arrays of elemental cells, each featuring identical arrangement of atoms over the large volumes of the crystal, is mostly unrealistic. Real crystals usually contain structural imperfections referred to as defects. High density of structural defects in semiconductor crystal prevents its use in the manufacture of high-performance devices because any departures from the lattice periodicity have an adverse effect on electrical characteristics of material, and hence, on the performance of semiconductors devices.

Depending on their nature, defects observed in semiconductor crystals fall into four classes which include point defects, line defects, planar defects, and volume defects. The most common point defects (Fig. 2.4(a)) result either from a vacancy formed by the atom missing from the lattice or from the excess atoms located interstitially (interstitial defect), or substitutionally (substitutional defect).

Line defects, or dislocations (Fig. 2.4(b)) can be viewed as a continuous array of point defects spreading throughout large portions of the crystal lattice. Two often encountered types of dislocations are edge dislocations, schematically represented in Fig. 2.4(b), in which dislocation

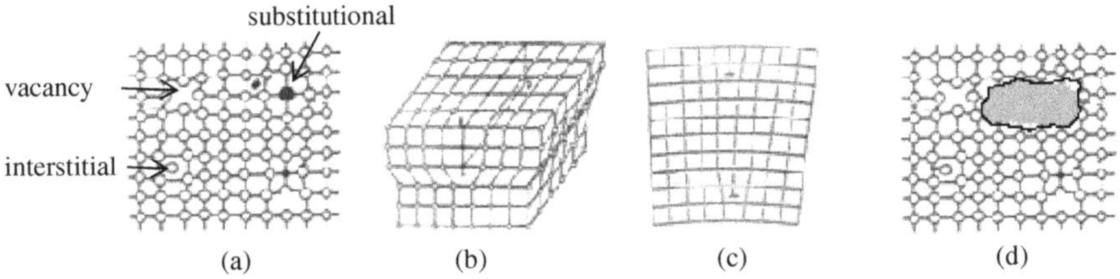

Fig. 2.4. Schematic representation of four types of structural defects occurring in single-crystal solids, (a) point defect, (b) line defect, (c) planar defect, and (d) volume defect.

line spread parallel to the stress in the lattice and screw dislocations in which dislocation line is perpendicular to the direction of the stress in the lattice. In real single-crystals often a mix of the edge and screw dislocations is observed.

Yet another class of crystallographic defects is represented by the planar defects. Also known as area defects (Fig. 2.4(c)), planar defects are basically the arrays of dislocations appearing in the crystal in the form of grain boundaries or stacking faults resulting from the disturbances in the single-crystal growth process. The last among defects that can be encountered in semiconductor single crystals are volume defects which are simply small-volume inclusions of non-crystalline phase into a single-crystal structure (Fig. 2.4(d)), or clusters of vacancies forming voids.

Engineering of the crystals towards minimization of the density of structural defects is a major objective in the processing of single-crystal materials intended for use in the manufacture of semiconductor devices. This is because in the case of high-density of structural defects in the single-crystal semiconductor, the adverse effect of defects may dominate over the intrinsic physical properties of the crystal.

2.3 Components of Semiconductor Device Material System

In order to fully exploit physical properties of semiconductors discussed in the previous Chapter, the device material systems comprised of elements featuring desired crystallographic structure and chemical composition need to be created. In this section parts comprising typical semiconductor device material system are identified. What needs to be emphasized at this point is that to perform desired functions that require

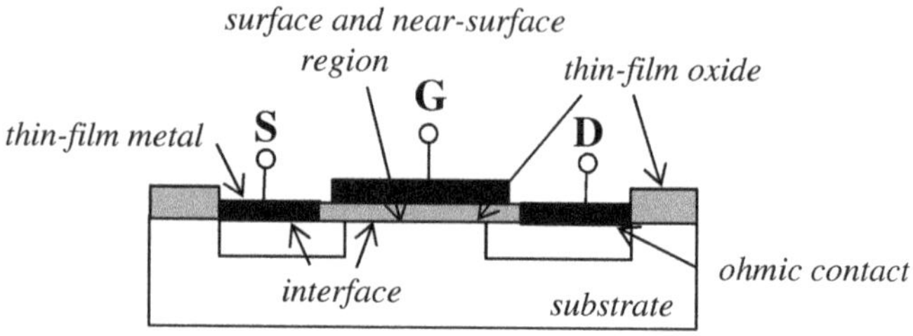

Fig. 2.5. Cross section of the MOSFET is used here to identify key components of a typical semiconductor device material system.

materials featuring semiconductor properties discussed Chapter 1, functional semiconductor device includes as well parts formed using conductors and insulators.

Figure 2.5 shows cross-section of the common transistor (metal–oxide–semiconductor field effect transistor (MOSFET)) which is used here to identify integral parts of essentially any semiconductor device. Parts comprising semiconductor device shown in this figure are introduced below and some are also discussed in more detail further in this book.

Substrate: Semiconductor device fabrication process starts with a substrate upon which other parts of the device are formed. In mainstream applications, substrate is a semiconductor material acting as an active part of the device while at the same time providing mechanical support for the device structure formed upon it. Alternatively, mechanical support for the devices formed on their surface, and thus acting as a substrate, can be provided by the materials which are not semiconductors, and which can be either electrically conductive or isolating, optically transparent or opaque, flexible or rigid.

In later parts of this book, issues concerned with substrates used in semiconductor device technology are discussed further.

Surface and interface: As the name indicates, surface is an exterior face of the solid and represents 2D termination of fundamental characteristics displayed by the three-dimensionally distributed atoms in the bulk of the sample (Irene, 2008). In the case of crystals, surface also represents an abrupt discontinuity of crystal structure. Furthermore, atoms on the surface, featuring at least one broken bond, are the only ones that can be exposed to the ambient. Such unsaturated bonds at the surface are

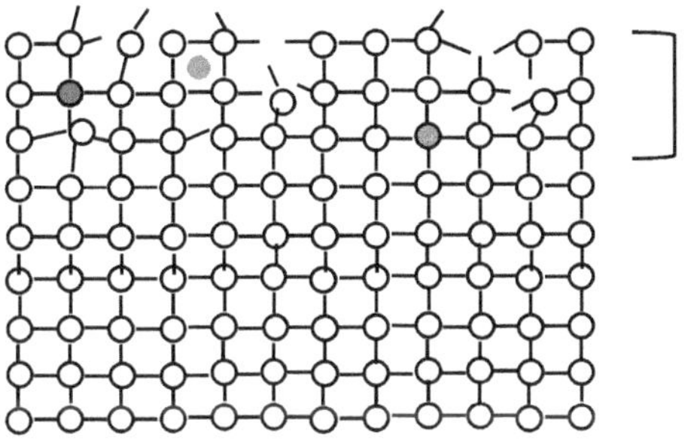

Fig. 2.6. Illustration of the state of the surface and near-surface region of single-crystal substrate.

electrically active and unless neutralized, these dangling bonds, commonly referred to as surface states, will feature electric charge altering distribution of electric charge in the sub-surface region of semiconductor. As a result, the impact of the surface on the properties of a solid is not limited to the single-atom plane at the surface but extends to the near-surface region (Fig. 2.6). In practice, such near-surface region can be significantly disturbed by surface roughness and process related physical damage.

Also, changes in the chemical characteristics of semiconductor surface affect magnitude and stability of electrical charge associated with any given surface chemistry which in turn has an effect on the distribution of electric charge in the near-surface region (Fig. 2.6). In practice, such near-surface region can be additionally disturbed by surface roughness and process related physical damage. Also, changes in the chemical characteristics of semiconductor surface affect magnitude and stability of electric charge in the near-surface region. It is a common practice to subject semiconductor surfaces to the treatments which result in the desired termination of broken chemical bonds on the very surface (surface termination) rendering such processed surface chemically passive (surface passivation). First and foremost, however, it is imperative that semiconductor surfaces in the course of device manufacturing processes are kept free from any contamination which would alter their physical and chemical characteristics. This is for that reason that surface cleaning operations are the most frequently applied processing steps in mainstream semiconductor device manufacturing (see discussion of surface processing in Chapter 5).

Considering all the above, electronic properties in the near-surface region of any solid, including semiconductors, depart significantly from

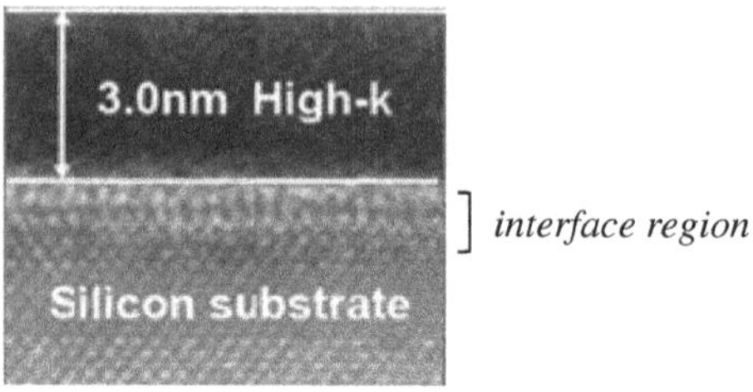

Fig. 2.7. Transmission Electron Microscopy (TEM) image of silicon substrate in the vicinity of interface with deposited high-*k* dielectric.

the same properties in the bulk (Fig. 2.7). For instance, due to the increased scattering of charge carriers resulting from the defective lattice and electrically charged centers in the near-surface region, mobility of charge carriers μ close to the surface is reduced significantly as compared to the bulk. As the result, following dependence of conductivity σ on carrier mobility μ discussed earlier, the electrical conductivity in the semiconductor sample is lower in its near-surface region than in the bulk. This effect has an adverse impact on the operation of semiconductor devices which depend on the surface and interface related phenomena (see discussion of field-effect transistors in Chapter 3).

Interfaces in turn are an integral part of any material system comprised of two or more materials where they play the role which defines characteristics of the entire system. As in the case of the surfaces, the effect of an interface between two materials expands into both adjacent regions. For the sake of illustration, Fig. 2.7 shows TEM micrograph of an interface between deposited ultra-thin amorphous dielectric (HfO_2) and single-crystal silicon (Chang *et al.*, 2007).

As figure shows, an interface is essentially a transition region featuring finite thickness that is needed to allow structural transition between two materials featuring different structure, and/or chemical transition between two materials featuring different chemical composition. In either case, interface represents a discontinuity of electrical, optical, mechanical, and thermal properties of the material system. From the point of view crystal structure, interface is basically a planar defect which is disrupting integrity of the material system, and hence, altering its characteristics. As the discussion in Chapter 3 will show, the impact of interface on semiconductor device performance manifests itself differently depending on whether device current is flowing parallel to the interface, or across the interface (Ruzyllo, 2025).

Thin films: Whether semiconductors, metals or insulators, thin films are an integral part of any semiconductor material systems where they play roles defined by their inherent characteristics. As discussed in Section 1.6, there is a correlation between physical dimensions of the piece of solid state material and its properties. While gradually confining geometry of a solid from the 3D bulk realm to the 2D configuration, basic physical characteristics of material are changing. The changes will occur according to the laws of classical physics up to the point where at extreme geometrical confinement (see Fig. 1.14) quantum phenomena take over control of the behavior of electrons which are now subject to the laws of quantum physics.

To illustrate this transition, Fig. 2.8 attempts to qualitatively present changes in the properties of semiconductor material as its thickness is gradually reduced using changes of semiconductor resistivity ρ with its thickness as an example.

Resistivity of semiconductor material displays bulk characteristics as long as continued reduction of the thickness does not alter electrons motion in any of the three directions. At certain thickness, however, resistivity starts increasing as the increasing scattering of strongly 2D confined electrons alters their flow. This is a point at which material no longer displays bulk properties and assumes properties of a thin film (Fig. 2.8).

Properties of thin films continue to change (increase of resistivity in Fig. 2.8) as the film thickness is decreasing. At certain thickness, typically in the range of few single atoms, the material transitions into the state of nano-confinement (see Section 1.6) in which laws of classical physics no longer apply and quantum phenomena begin to dominate physical

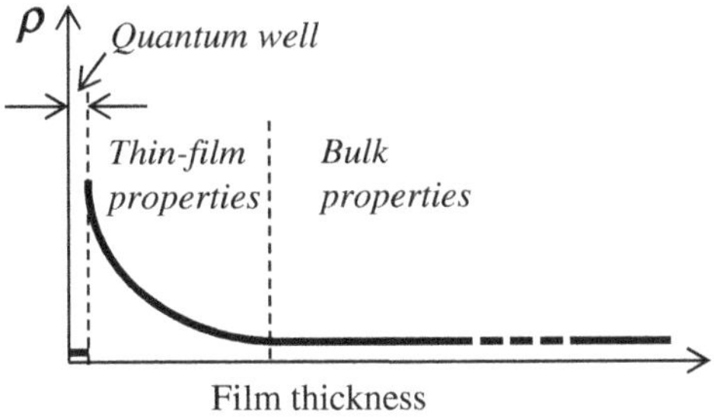

Fig. 2.8. Simplified qualitative illustration of the change in semiconductor material resistivity ρ as its thickness is decreasing.

properties of the material. In this state, electrical resistivity ρ has no effect on electrons flow as the scattering-free ballistic transport controls the motion of charge carriers in 2D material. This transport is possible in two dimensions only as the electrons are strongly confined in the third dimension. The corresponding condition is referred to as 2D electron gas (2DEG) which is a condition characteristic of electrons behavior in 2D material systems known as quantum wells ("wells" because of lower potential within 2DEG part of the device as compared to the potential in the adjacent parts). The unusual and advantageous properties of 2DEG are often exploited in advanced semiconductor devices whenever increase of electrons' velocity is needed to speed up device operation.

It needs to be pointed out that the universally valid number defining thickness below which thin film rather than bulk properties define characteristics of material, cannot be given because the thickness at which transition from the bulk-dominated properties to thin-film properties occurs is different for different materials. In addition, it depends on the crystallographic structure of material, its purity, and defect density.

Ohmic contacts: The integral components of any semiconductor material system processed into a functional device are ohmic contacts through which electrical communication between the device and external electronic circuitry is established (Fig. 2.5). The term *ohmic* is used in reference to contacts between conductive materials, in the case of semiconductor devices typically between metal and semiconductor. Ohmic contacts are featuring very low, independent of applied voltage series resistance R_s. As a result, current–voltage (I–V) characteristic of ohmic contact is linear, symmetric, and features high slope reflecting very low series resistance of the contact ($R_s \rightarrow 0$) as shown in Fig. 2.9.

As an example of applications of ohmic contacts, Fig. 2.5 shows source and drain ohmic contacts in the conventional transistor which to operate properly, needs to feature I–V characteristic as shown in Fig. 2.9. The same ohmic contacts related considerations apply to any semiconductor device as in one way or another, all semiconductor devices need to be equipped with contacts assuring undisturbed flow of electric current in and out of the device.

In general, in the selection of materials forming ohmic contacts with any given semiconductor as well as in the processing of such contacts needs to be carefully considered. The metal forming contact to

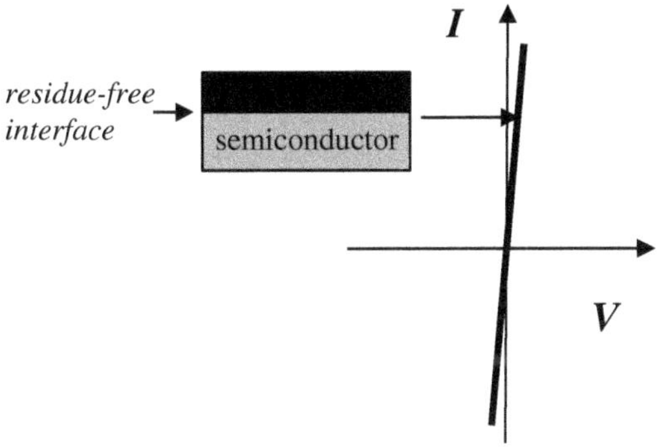

Fig. 2.9. Metal-semiconductor ohmic contact and its current–voltage (I–V) characteristic.

semiconductor material must be selected such that there is no potential barrier formed at the interface (Ruzyllo, 2025). Furthermore, ohmic contacts technology involves preparation of the semiconductor surface prior to metal deposition as discussed in Chapter 6.

In summary, discussion in this section leads to the conclusion that each element of the semiconductor device material system exemplified by the transistor in Fig. 2.5, plays a role in defining characteristics of functional semiconductor devices. The discussion in Chapter 3 shows correlation of the effects discussed above with the operation of actual semiconductor devices.

2.4 Inorganic Semiconductor Materials

Semiconductor materials can be divided into various classes based on the range of criteria (Tyagi, 1991; Levinshtein, 1999; Madelung, 2004; Rockett, 2008; Yacobi, 2008). In this chapter semiconductors are considered in terms of their chemical composition distinguishing between inorganic semiconductors and organic semiconductors. Among the former, elemental semiconductor materials and compound semiconductor materials are considered as the distinct classes. First, we consider inorganic semiconductors understood here as the solids which do not contain carbon–hydrogen (C–H) bonds in their structure. Inorganic semiconductors represent a dominant class of semiconductor materials upon which broadly understood semiconductor technology was developed over the years (Ruzyllo, 1998).

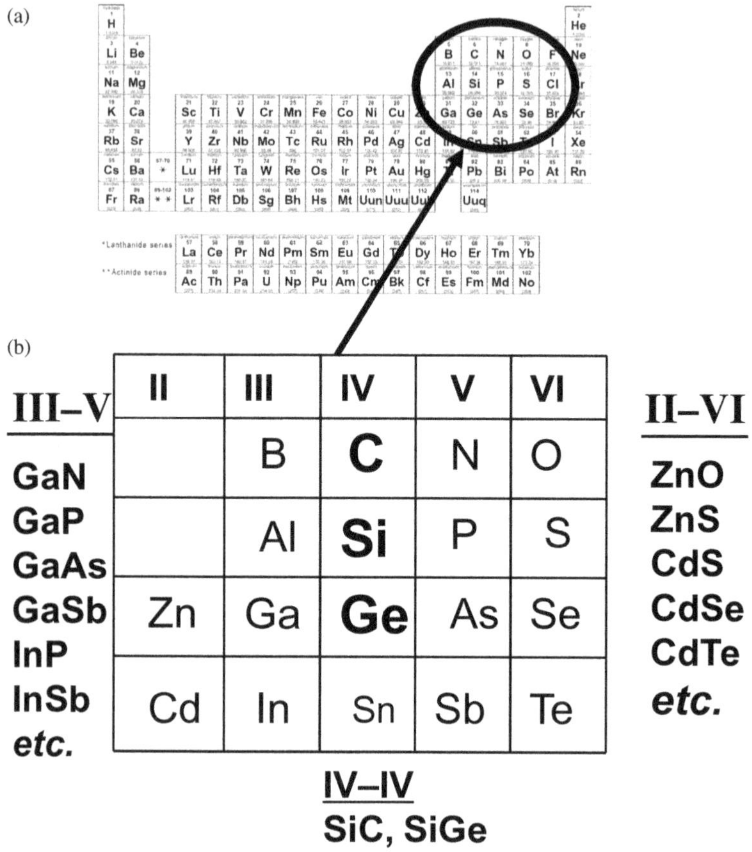

Fig. 2.10. (a) Periodic Table of Elements and (b) part of it referred to as semiconductor periodic table from which elemental and compound semiconductor materials originate.

In the Periodic Table of the Elements shown in Fig. 2.10(a), a section singled out in Fig. 2.10(b) is often referred to as a "semiconductor periodic table". All inorganic semiconductor materials either elemental in group IV such as silicon (Si), or in the form of compounds synthesized using elements from the group IV (IV–IV compounds, e.g. silicon carbide, SiC), from the groups III and V (III–V compounds, e.g. gallium nitride, GaN), as well as from the groups II and VI (II–VI compounds, e.g. cadmium selenide, CdSe), originate from this section of the periodic table.

2.4.1 *Elemental Semiconductor Materials*

Elements from group IV of the periodic table (Fig 2.10(b)) feature four valence electrons and include carbon (C), silicon (Si), germanium (Ge), and tin (Sn). Among them tin, with energy gap E_g close to zero, displays very weak semiconductor properties and in addition features melting point of just 223 °C. Hence, as an element, tin is not used to manufacture semiconductor devices. Key characteristics of the remaining three semiconductor elements are reviewed as follows.

Selected physical characteristics of the three elemental semiconductors are summarized in Table 2.1. Values shown in this Table, as well as in other Tables listing parameters of various semiconductors in this chapter are tabulated for the purpose of comparison between various materials and may not reflect accurate numbers that could be used as a reference.

Carbon (C): In addition to silicon and germanium, carbon is the third element in group IV of the Periodic Table (Fig. 2.10(b)) which in selected structural configurations displays well defined semiconductor properties. In nature, carbon most commonly appears in the form of graphite. Graphite is a crystal featuring hexagonal structure which with a

Table 2.1. Selected characteristics of single-crystal elemental semiconductor materials.

	Silicon, Si	Germanium, Ge	Carbon, C (diamond)
Energy gap width E_g (eV)	1.1	0.77	5.45
Energy gap type	indirect	indirect	indirect
Electron mobility at 300K (cm²/Vs)	1500	3900	2200
Hole mobility at 300K (cm²/Vs)	450	1900	1600
Electron Saturation velocity (cm/s)	2.7×10^7	7×10^6	2×10^7
Saturation velocity el. field (V/cm)	2×10^4	2×10^3	10^5
Breakdown field (V/cm)	3×10^5	10^5	5×10^6
Thermal conductivity (W/cmK)	1.4	0.6	22
Thermal expansion coeff. (K^{-1})	4.05×10^{-6}	5.8×10^{-6}	0.8×10^{-6}
Melting point (°C)	1414	938	sublimation

Note: Approximate values are shown for comparison purpose.

potentially attractive electron mobility of around 3,000 cm^2/Vs, but very narrow energy gap E_g of less than 0.05 eV displays electrical characteristics of a conductor rather than that of semiconductor. Similarly, the most fundamental single atom thick component of the graphite structure known as graphene, while displaying significantly higher than graphite electron mobility, features no naturally occurring energy gap which limits its applications in the mainstream semiconductor device engineering (see further discussion in this section).

In the 3D single-crystal configuration carbon crystallizes in the form of a diamond. Diamond crystal lattice belongs to the cubic class of crystals (Fig. 2.2) and represents the same crystallographic pattern which is followed by the arrangement of atoms in single-crystal silicon. In contrast to graphene which lacks energy gap, diamond has a very wide energy gap $E_g = 5.4$ eV (Table 2.1) making it a key material in the wide-bandgap class of semiconductors with great potential in the manufacture of the devices designed to operate at high power and elevated temperature. In addition, as Table 2.1 indicates, diamond features very good charge carriers' mobility characteristics with holes mobility significantly higher than the mobility of holes in silicon and similar to the mobility of holes in germanium in the case of formed at high-pressure, high-temperature bulk diamond as well as epitaxially deposited thin-film diamond (Grot *et al.*, 1991). The most outstanding characteristic of diamond, however, is its very high thermal conductivity of 22 W/cmK, the highest among all semiconductor materials. The resulting diamond's excellent heat dissipation characteristics are of great importance in the case of devices which by the nature of the electronic functions they perform generate large amounts of heat.

The reason why despite several highly advantageous characteristics listed above diamond is not finding broad applications in the manufacture of semiconductor devices is related to the doping challenges. Due to its fundamental properties, as processed diamond is a *p*-type semiconductor material and because of the lack of dopants able to efficiently convert it into *n*-type semiconductor, physical characteristics of diamond cannot be fully exploited in practical devices. Also, the size and crystal quality of diamond attainable with economically viable technology are not compatible with the needs of commercial mass manufacturing of semiconductor devices. An alternative to the bulk-diamond technology could be provided by the continuously improving thin-film nanocrystalline diamond deposition technology, although doping challenges will remain valid.

Among stand-alone 2D nanoscale materials, one carbon atom thick graphene is particularly well explored (Aoki and Dresselhaus, 2014). In terms of crystallographic structure, graphene is a 2D part of the 3D graphite. It is a one atom thick planar sheet of carbon that is connected to adjacent sheets forming graphite by relatively weak van der Waals forces (Bonca, 2010). Effectively then, graphite is a thick stack of graphene sheets each internally covalently bonded (see Section 1.3) in the planar hexagonal configuration, but which at the same time are relatively weakly bonded between each other (Fig. 2.11(a)). Rolled into a cylinder, graphene forms a single-walled carbon nanotube, CNT, (Fig. 2.11(b)) which forms another unique carbon nanoscale configuration, and which is discussed later in this section.

One of the remarkable characteristics of graphene is its extremely high electron mobility reported in the range from 150000 to 200000 cm^2/Vs and accounting for its excellent electrical conductivity. Interestingly, observations based on the conductance measurements suggest that in graphene electrons and holes mobilities are almost the same. However, the numbers regarding carriers mobilities are significantly reduced when graphene is brought into contact with other materials. Unfortunately, contact with other materials is unavoidable when graphene becomes an active part of the functional device.

The use of graphene in certain types of semiconductor devices designed to carry out electronic functions, e.g. transistors in the logic integrated circuits (see Chapter 4), is limited because graphene features lack of the energy gap needed to efficiently execute on-off functions.

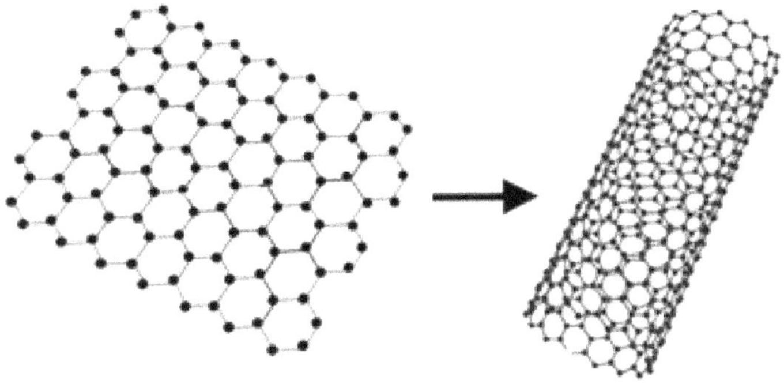

(a) (b)

Fig. 2.11. (a) Single-atom thick graphene is rolled up into (b) single-walled carbon nanotube, CNT.

Without a bandgap, graphene is essentially a conductor which cannot be turned off completely, and thus, in its pure-carbon form (i.e. without opening of the bandgap by means of manipulations of its chemical makeup) cannot be used to make transistors for logic applications. This limitation does not preclude exploitation of the outstanding characteristics of graphene in other semiconductor device related uses.

Silicon (Si): Silicon is the most important and the most widely used semiconductor material (Wolf, 1969; Wolf and Tauber, 1986). Its use continues to grow through the needs of the well-established as well emerging device technologies (see Chapter 3). What makes silicon so special is a combination of abundance (after oxygen, silicon is the second most common element in the Earth crust), low cost of obtaining high-quality crystals as compared to other key semiconductors, and overall characteristics making it compatible with the needs of mass commercial manufacture of semiconductor devices.

As data in Table 2.1 indicates, intrinsic physical characteristics of silicon are not necessarily its strength. Low mobility of holes in silicon as compared to germanium and carbon makes implementation of silicon devices in which holes transport is a factor is a challenge. The indirect energy gap renders silicon ineffective in the conventional light generation applications, but does not eliminate silicon from the light generation applications.

On the positive side, an outstanding characteristic of silicon with far reaching consequences in terms of how it can be used in device manufacturing is the fact that the native oxide of silicon, silicon dioxide SiO_2, is an excellent insulator. No other semiconductor material has a native oxide of such high quality. This characteristic put silicon in the class on its own in this respect. Also, excellent mechanical properties of silicon, elastic properties in particular, make Si highly suitable for electro-mechanical device applications (see discussion of MEMS devices in Chapter 3). Excellent elastic properties mean that after being deformed or flexed, silicon very quickly and with negligible energy dissipation regains its original shape. In the case of silicon this cycle can be repeated almost endlessly without material fatigue allowing for highly repeatable motion without breaking virtually indefinitely.

Silicon in the form of crystal is at room temperature brittle, but otherwise mechanically sturdy, relatively high melting point material (see Table 2.1) which meets the challenges of the semiconductor device manufacturing including exposure to high temperatures and extensive robotic handling (Hull, 1999). Another worthy feature of silicon is that it can be readily obtained in any crystallographic form including single-crystal, polycrystalline, and amorphous phases. In terms of the quality and size of the crystal, single-crystal Si in the form of circular, or rectangular depending on application wafers, are not matched by any other semiconductor. In contrast to other semiconductor materials, single-crystal Si can be processed into wafers as large as 450 mm in diameter and potentially larger. Also, commercially available are large rectangular wafers of multicrystalline silicon finding use in solar cells applications. Very common in the range of applications are thin films of amorphous silicon fabrication technology of which is very well established.

Availability of dopants which can make any given semiconductor material either *n*-type or *p*-type (see Section 1.3) is a property defining its usefulness in the manufacture of functional semiconductor devices. In the case of silicon, group III boron (B) serves as an effective acceptor, or *p*-type dopant, while elements from the group V, phosphorus (P), arsenic (As) and antimony (Sb) can be selected (Fig. 2.10) depending on the peculiarity of the process, to act as *n*-type dopants.

Furthermore, good thermal conductivity and advantageous high electric field characteristics of silicon (see saturation velocity in Table 2.1) contribute to the versatility of silicon as a semiconductor material. However, this is a broad range of intangible, manufacturability and cost related characteristics of silicon that decide on the dominant role it plays in semiconductor device engineering.

Silicon is also used in selected nanoscale configurations presented in Fig. 1.15. In 2D configuration it is applied in the form of nanosheets acting as channels in transistors in logic integrated circuits. In 1D configuration in turn, it can be used as a nanowire. Similarly to carbon nanotubes (CNT), also in the case of silicon nanowires the nanoconfinement brings about drastic modifications of the physical characteristics of silicon as compared to its bulk properties. As a result, silicon in the form of nanowires display electrical, optical, and mechanical characteristics which open areas of applications not available to silicon in the bulk form. The most significant effect of silicon confinement into the nanowire is dependence of its bandgap width on the diameter of the wire. The energy gap E_g of silicon can potentially be increased to above 2 eV (E_g of bulk Si is 1.1 eV) in the case diameter of silicon nanowires (SiNW) could be reduced to around 2 nm. Equally significantly, the type of energy gap can be converted from indirect to direct which allows silicon to be used in light emission devices.

A unique feature not only of SiNWs, but nanowires in general is a very large surface-to-volume ratio making surface characteristics play a role in defining properties of the 1D nanoscaled materials. In the case of silicon nanowire, the way its surface is terminated/passivated is a factor controlling basic characteristic of the nanowire.

Silicon nanowires can be fabricated depending on the needs using various methods which belong to the broadly defined classes of either top-down or bottom-up techniques (see Chapter 5). Without getting into the details of the mechanisms involved in the process of SiNWs formation, let it be noted that the end results of the former are nanowires positioned on the substrate surface horizontally (Fig. 2.12(a)), while the latter will typically result in nanowires formed on the surface vertically (Fig. 2.12(b)). Although these are the same silicon NWs, some of their characteristics change depending on how they were formed and configured on the substrate surface. Allowing some simplifications, it can be said that the horizontally positioned SiNWs are mostly compatible with the needs of electronic (flow of the electric current) devices, while densely packed, vertically configured SiNWs are featuring characteristics conducive with the needs of selected photonic devices.

Considering all of the above there is no doubt that nanowires in general, and silicon nanowires in particular, will find in the future various applications in semiconductor devices. It is recommended that the basic characteristics of semiconductor nanowires are kept in mind while reviewing Chapter 4 concerned with semiconductor devices.

In summary, the key properties of silicon considered above make it the most important and the most broadly used semiconductor material. Considering directions in which semiconductor technology grows and expands, its leading role will not be challenged in the foreseeable future.

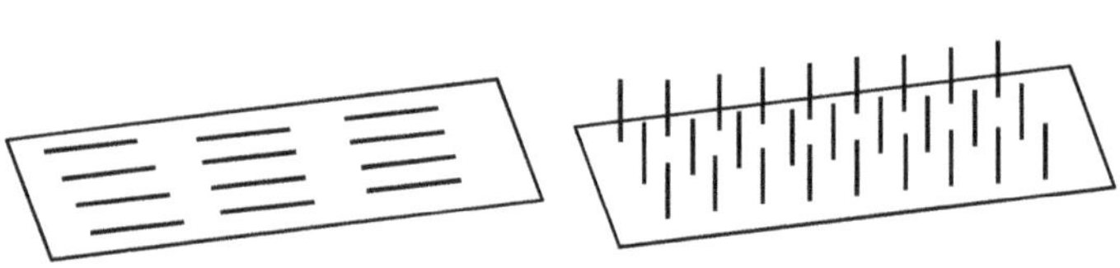

(a) (b)

Fig. 2.12. Schematic representation of (a) horizontal, (b) vertical SiNW.

Germanium (Ge): Before silicon took over as a lead semiconductor material, germanium was semiconductor of choice in very early device applications. In fact, the very first transistor constructed in 1947 was using germanium as semiconductor material. The key deficiency of germanium as compared to silicon is the fact that its native oxide, germanium dioxide GeO_2 is unstable and decomposes under the normal conditions including exposure to water.

Germanium itself is also less chemically resistant and mechanically stable than silicon. Also, higher cost of high-quality single-crystal germanium as compared to silicon is the factor limiting broad use of germanium in commercial applications. On the other hand, however, higher charge carriers mobilities, holes in particular, than in silicon (Table 2.1) support viability of germanium in various specialized device applications. In terms of doping, or in other words making germanium either *n*-type or *p*-type, the same elements from the group III and group V respectively as in the case of silicon (see earlier discussion) can be used. In spite of some advantageous features, overall characteristics of germanium make it less common than silicon in semiconductor device manufacture.

2.4.2 *Compound Semiconductor Materials*

Besides elemental semiconductors, materials displaying strong semiconductor properties can be synthesized using elements from the groups II to VI in the semiconductor periodic table shown in Fig. 2.10. Unlike occurring in nature elemental semiconductors, compound semiconductors used to manufacture commercial devices are all manmade which makes a material engineering element of the compound semiconductor device technology to be particularly important (McGuire and Strausser, 2010).

Various types of compound semiconductors discussed in the following are grouped based on the position in the periodic table of the elements comprising a compound. Specifically, as an example, compounds synthesized using different elements from the group IV (IV–IV compounds), compounds formed using elements from groups III and V (III–V compounds), and compounds formed using elements from groups II and VI (II–VI compounds) are distinguished and in this order considered below.

IV–IV Semiconductor Compounds: The IV–IV class of semiconductor compounds (Fig. 2.10) is represented by silicon compounds with carbon (silicon carbide, SiC) and with germanium (silicon germanium, SiGe).

As the discussion below will indicate both compounds, each in its own way contributes meaningfully to semiconductor device engineering.

***Silicon Carbide (SiC)*:** Also known as carborundum, SiC exists in over 100 different polytypes (families of crystals displaying similar structure) which vary in the details of the long-range stacking order within the crystal. In terms of two broad families of crystals, SiC occurs in hexagonal and cubic configurations. In the former case two polytypes 4H-SiC (β-SiC) and 6H-SiC (α-SiC) are the most common, while cubic polytypes are most often represented in semiconductor technology by 3C-SiC polytype. Choices regarding type of polytype used depend on the specific device application.

Each SiC polytype features somewhat different properties. The values quoted below are for the 4H-SiC (β-SiC) polytype and are used here as representative of the entire class of SiC crystals. First and foremost, SiC is a wide-bandgap semiconductor featuring indirect energy gap $E_g = 3.2$ eV. Mobilities of charge carriers in β-SiC are 900 cm^2/Vs and 100 cm^2/Vs for electrons and holes respectively. Thermal conductivity of 4.9 W/cmK, although not as high as that of the diamond (Table 2.1), is higher than in elemental semiconductors. It features high breakdown field of 3×10^6 V/cm, and high saturated electron drift velocity in the range of 10^7 cm/s. This combination of characteristics offers superior performance of SiC devices in high temperature/high power applications, as well as in high speed operation under the high electric field stress. In other words, SiC is a material allowing devices operating under particularly strenuous conditions in terms of temperature (even above 500 °C), high currents, and voltages, as well as high electric field.

In terms of processing, different characteristics of Si and C comprising SiC, causing different responses to the process ambient are coming to play (Fig. 2.13(a)). In the course of thermal oxidation, for instance, gaseous carbon oxides are escaping SiC and the oxide remaining on its surface is silicon dioxide SiO_2. The process leaves the surface region of SiC deprived of oxygen atoms (Fig. 2.13(b)) which manifest itself in characteristic for SiC surface feature after thermal oxide is etched off (Fig. 2.13(c)).

Thermal oxide grown on SiC surface oxide is electrically sufficiently stable to make this semiconductor material suitable for the manufacture of the metal–oxide–semiconductor (MOS) based devices discussed in Chapter 3. Also other fabrication processes are mostly compatible with

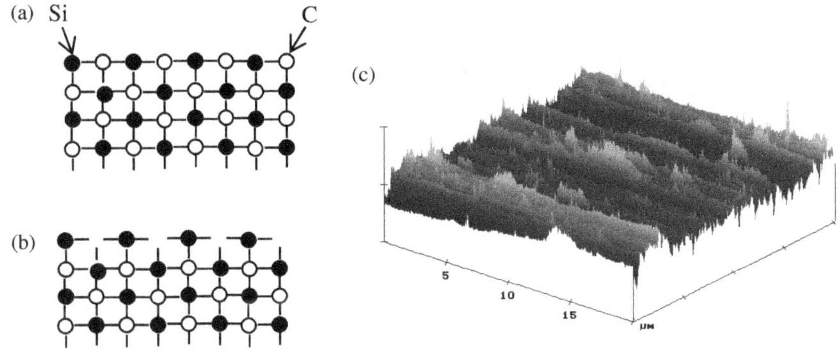

Fig. 2.13. Surface features of a binary semiconductor compound formed using elements Si and C displaying different oxidation characteristics (a) before thermal oxidation (b) after oxidation, and (c) AFM after thermal oxidation followed by oxide etching.

processing of silicon carbide (Luther *et al.*, 1993). On the other hand, due to the indirect bandgap SiC cannot be used in the light-emitting device applications in which its wide bandgap could be put to good use. In terms of dopants, *p*-type and *n*-type doping and conductivity control can be achieved by doping with aluminum and nitrogen respectively. Silicon carbide is colorless and transparent to visible light.

In the light of its advantageous characteristics, SiC is broadly pursued commercially wide-bandgap semiconductor material in power devices. Availability of high quality 200 mm single-crystal SiC wafers assures growth of silicon carbide technology in industrial endeavors.

***Silicon germanium (SiGe)*:** Another IV–IV semiconductor compound based on silicon is silicon germanium (Fig. 2.10). In contrast to SiC, silicon germanium does not exist in the defined lattice, but rather as a mixed crystal properties of which, including width of the energy gap and carrier mobility, can be modified by changing ratio of Si and Ge atoms in the mix. At 1:1 composition, SiGe features energy gap narrower than Si and wider than Ge, and electron mobility higher than in silicon and lower than in germanium.

The result of mixing Si and Ge atoms and forming SiGe is that the lattice constant of the resulting and is different than that of silicon. Taking advantage of this feature, SiGe is commonly used to introduce strain into Si lattice by acting as a substrate on which thin films of single-crystal silicon are deposited. As indicated in Chapter 1, strain in the crystal lattice is increasing electron mobility. Hence, SiGe is used to introduce strain in silicon by which improved performance of selected devices can be achieved (see discussion in Chapter 3).

***Germanium-tin (GeSn)*:** While not used as an elemental semiconductor, tin alloyed with germanium results in the compound offering interesting semiconductor properties. Simulation and experiments indicate that by forming a germanium-tin alloy, GeSn, with 3% of Sn added to Ge, a semiconductor featuring direct bandgap, as opposed to indirect gap of germanium, and increased hole mobility as compared to that of Ge is obtained.

III–V Semiconductor Compounds: The III–V compounds are typically classified based on the group V element forming a compound (Fig. 2.10) into nitrides, phosphides, arsenides and antimonides. A review below follows this classification.

***Nitrides*:** In this class of compound semiconductors gallium nitride, GaN, is of key importance in both electronic and photonic devices. Outstanding characteristics of GaN is its direct and wide (E_g = 3.5 eV) energy gap (Table 2.2). These bandgap's characteristics make GaN uniquely suitable for the emission of short wavelength light in the blue range. As a result,

Table 2.2. Characteristics of selected direct-bandgap single-crystal III–V compounds.

	Gallium nitride GaN	Indium phosphide InP	Gallium arsenide GaAs	Galium antimonide GaSb
Energy gap width E_g (eV)	3.5	1.35	1.43	0.72
Electron mobility at 300K (cm^2/Vs)	1,000	5.400	8.500	3.000
Hole mobility at 300K (cm^2/Vs)	350	200	400	1,000
Electron Saturation velocity (cm/s)	2×10^7	7×10^6	2×10^7	8×10^6
Breakdown field (V/cm)	5×10^6	5×10^5	4×10^5	4×10^4
Thermal conductivity (W/cmK)	1.3	0.68	0.55	0.32
Lattice	Hexagonal	Cubic	Cubic	Cubic
Lattice constant (nm)	0.316/0.512	0.587	0.565	0.609

Note: The values shown are for comparison purposes only.

and in addition to the lack of other single-crystal semiconductors featuring similar characteristics, GaN is a cornerstone material in blue light emitting semiconductor devices. Furthermore, wide bandgap makes it highly suitable for high-power/high-temperature device applications.

In the light of limited availability of freestanding, low-cost, large-area single-crystal GaN substrates, GaN devices are often fabricated using thin-film GaN deposited on the substrates made out of the other materials including sapphire, silicon carbide, and silicon.

Among other III–V nitrides, boron nitrid (BN) and aluminum nitride (AlN) are attracting attention due to the largest, direct energy gap among all semiconductor compounds (BN E_g = 6.4 eV, AlN E_g = 6.2 eV). However, at the low electron mobility (BN μ = 200 cm^2/Vs, AlN μ = 300 cm^2/Vs) their use in commercial electronic devices is limited primarily to UV detection. Indium nitride (InN) on the other hand features rather narrow bandgap ($E_g \sim 0.7$ eV) and relatively high electron mobility μ = 3,200 cm^2/Vs. In spite of it, InN is best used when alloyed with GaN to form InGaN.

Phosphides: Three among III–V phosphides namely boron phosphide (BP), aluminum phosphide (AlP), and gallium phosphide (GaP) all feature relatively wide, but indirect bandgap E_g = 2.1 eV, E_g = 2.5 eV, and E_g = 2.26 eV, respectively, and low electron mobility of 500 cm^2/Vs, 80 cm^2/Vs and 110 cm^2/Vs respectively. Negative feature of aluminum phosphide is its high toxicity while positive characteristic gallium phosphide is that with small amount of Al added the energy gap of GaP can be converted from indirect to direct. In this context, indium phosphide, InP distinguishes itself among III–V phosphides with a direct bandgap (E_g = 1.35 eV) and decidedly higher than other phosphides electron mobility of 4,500 cm^2/Vs (Table 2.2). Combination of these features makes InP useful as a material for infrared emitters and detectors as well as in high-speed electronics.

Arsenides: Among III–V arsenides a difference in the key properties between aluminum arsenide (AlAs), and boron arsenide (BAs), on one hand and gallium arsenide (GaAs), and indium arsenide (InAs) on the other is noted. The former two feature relatively wide, indirect bandgap (AlAs, E_g = 2.2 eV, BAs E_g = 1.5 eV) and low electron mobility of 200 cm^2/Vs for AlAs (much higher in 2D configuration) which is used primarily as a component in ternary III–V alloys such as AlGaAs. On the other hand, distinguishing feature of the cubic boron arsenide (BAs) is exceptionally high thermal conductivity. Its electron mobility is in the range of 1,500 cm^2/Vs.

Two remaining arsenides, namely gallium arsenide (GaAs), and indium arsenide (InAs) are important contributors to semiconductor device technology. In particular, GaAs is the best developed III–V semiconductor compound long exploited in the range of electronic and photonic semiconductor device applications (Baca and Ashby, 2005). It features direct, relatively wide (E_g = 1.43 eV) energy gap making it of importance in photonic applications. In addition to the direct bandgap, GaAs features high electron mobility of 8,500 cm^2/Vs which makes it also a key material in the range of high-speed electronic devices. The bandgap of GaAs can be effectively engineered by forming ternary compounds through addition of aluminum (AlGaAs). On the negative side, GaAs does not form on its surface good quality native oxide which prevents its use in metal-oxide-semiconductor (MOS) devices (see Chapter 3).

A distinct feature of GaAs is significantly higher vapor pressure of arsenic as compared to gallium. When exposed to temperatures above 600 °C, arsenic evaporates leaving behind gallium which collapses into the powder-like material. At the reduced pressure conditions, even at the lower temperature, evaporation of arsenic from the near-surface region can be observed leaving the surface of GaAs crystal arsenic deficient (similarly to missing carbon in Fig. 2.13(b)).

The remaining arsenide, indium arsenide, InAs, features narrower as compared to GaAs direct bandgap (E_g = 0.36 eV), but significantly higher electron mobility μ_n in excess of 20,000 cm^2/Vs and is an established III–V semiconductor compound in commercial devices. Furthermore, indium arsenide forms a ternary alloy with GaAs (InGaAs) which similarly to AlGaAs finds variety of electronic and photonic device applications.

Antimonides: Among antimonides, aluminum antimonide, AlSb and boron antimonide, BSb feature indirect bandgap (E_g = 1.6 eV and 0.5 eV respectively) and low carrier mobility. For that reason, both AlSb and BSb are not viable candidates for the broad commercial device applications. In contrast, remaining two antimonide semiconductors, gallium antimonide, GaSb (E_g = 0.72 eV) and indium antimonide InSb (E_g = 0.17 eV) both feature direct energy gap and attractive carrier mobility characteristics. In the former case not only high electron mobility of 5,000 cm^2/Vs attracts attention, but even more so, relatively high hole mobility (850 cm^2/Vs) makes GaSb of interest in certain device applications. Indium antimonide, InSb features the narrowest bandgap ($E_g = 0.17$ eV) among practical semiconductors, but at the same time very high electron mobility in the range of 80,000 cm^2/Vs. While very high electron mobility would make InSb attractive in electronic applications, its narrow energy gap focuses its usefulness in the manufacture of commercial photonic devices involving detection of mid-infrared radiation.

Overall, III–V compounds constitute an important class of semiconductor materials in both electronic and photonic applications. Several of them feature direct energy gap, and all feature either high carrier mobility, or wide bandgap, or both.

Ternary and quaternary III–V semiconductor compounds: Up to this point a review of III–V compounds was concerned with binary III–V compounds such as GaAs. In practice, two-element III–V compounds are often expanded into ternary alloys involving three elements from groups III and V.

There are several reasons for the inclusion of additional elements into III–V binary compound. First, is the device engineering driven need for either gradual changes of its bandgap E_g (bandgap engineering) or its crystal lattice constant *a* (Fig. 2.2), or both. The former modifies the spectrum of light which can be emitted and absorbed by the compound while the latter defines characteristics of the crystal lattice conducive with the needs of lattice-matching in the formation of the complete material systems. Taking aluminum gallium arsenide (AlGaAs) as an example, we can see that by changing Al fraction *x* in $Al_xGa_{1-x}As$, material transitions from gallium arsenide GaAs ($x = 0$) to aluminum arsenide AlAs ($x = 1$). In the process, the bandgap of the compound changes from $E_g = 1.42$ eV for GaAs to $E_g = 2.16$ eV for AlAs with negligible changes in the lattice constant.

Besides changes of the width of the energy gap, also type of the gap can be converted from direct to indirect (or *vice versa*) as a result of modifications in the ternary compounds chemical composition. For instance, direct bandgap InP can be converted into indirect bandgap GaP by gradually changing In fraction *x* in $In_xGa_{1-x}P$. In addition to the bandgap and lattice constant *a* engineering, changes in chemical composition from the binary to ternary alloy also modify its optical properties manifested in the changes of refractive index *n* which is a feature often exploited in the construction of semiconductor photonic devices.

To accomplish any of the goals listed above even more effectively, an option of adding fourth element into the ternary III–V compound is often exploited. For instance, ternary compound AlGaAs can be modified toward desired characteristics of the bandgap by forming a quaternary compound aluminum gallium arsenide phosphide, AlGaAsP.

While modifications of chemical composition of III–V compounds using elements from either group III or group V adjust characteristics of material in the ways discussed above, addition of the elements from other groups of the Periodic Table may alter fundamental characteristics of III–V compound in a more profound manner. For instance, by adding manganese (Mn), originally non-magnetic GaAs will acquire well defined ferromagnetic properties, and hence, will turn into a magnetic semiconductor GaMnAs featuring significantly increased magnetic susceptibility. This type of semiconductor material engineering finds its application in broadly understood spintronics.

II–VI Semiconductor Compounds: In addition to elemental semiconductors in group IV of the Periodic Table and compounds formed using elements from groups III and V, also elements from groups II and VI (Fig. 2.10) can be synthesized into II–VI compounds representing yet another distinct class of inorganic semiconductors. They are briefly reviewed below following classification based on the group VI elements of the compounds which considers them in terms of oxides, selenides, sulfides, and tellurides.

From among group II elements contributing to the binary II–VI compounds, namely zinc (Zn), cadmium (Cd), and mercury (Hg), the last one does not form binary compounds in any combination with group VI elements that would be of interest in semiconductor device applications. Therefore, Hg-based binary II–VI compounds are considered in the overview below only as the components of the ternary and quaternary II–VI compounds.

It should be also pointed out that several among II–VI compounds crystallize in more than one crystallographic structure, for instance, cubic or hexagonal (Fig. 2.2), depending on the conditions of the crystallization process. Considerations of the complex issues concerned with the peculiarities of the II–VI semiconductor compounds crystallography are beyond the scope of the overview below.

***Oxides*:** Among II–VI oxide semiconductors, zinc oxide (ZnO) is of greatest practical importance. It features direct, relatively wide bandgap ($E_g = 3.3$ eV) which makes it suitable for short wavelength blue and violet light emission (see Chapter 3). It is also highly transparent to white light, and thus can be used to construct transparent electronic and photonic elements. In general, it features somewhat higher than other II–VI compounds electron mobility which makes it better aligned than other II–VI compounds with the needs of electronic devices. Overall, ZnO can be seen as an II–VI alternative to the III–V GaN in electronic and photonic applications.

The other II–VI oxide semiconductor of interest in select electronic and photonic applications is gallium oxide Ga_2O_3. With energy gap E_g in the range of 4.7 eV it shows promise, despite relatively low thermal conductivity, in high-power electronic as well as UV detecting devices.

***Selenides*:** Both II–VI selenides, zinc selenide (ZnSe), $E_g = 2.7$ eV, and cadmium selenide (CdSe), $E_g = 1.75$ eV, feature direct bandgap. The former is used for blue light emission and UV detection. Due to the wide transmission wavelength range it is also used in IR (infrared) optics. Cadmium selenide (CdSe) is useful in various photonic applications. It is also used in the form of 0D quantum dots in which wavelength of emitted light can be controlled by changing diameter of the dot enforcing changes in the width of the energy gap which in standard 3D configuration and at room temperature is 1.75 eV.

***Sulfides*:** Among II–VI sulfides zinc sulfide (ZnS) crystal features the widest among all semiconductors energy gap $E_g = 3.6$ eV. With that wide, direct energy gap, ZnS is a semiconductor uniquely predisposed for applications in short-wavelength (blue and UV) light emission and detection. Similarly, cadmium sulfide, CdS, $E_g = 2.42$ eV is equally useful in light detection application. In addition, it is commonly used as a component of the CdTe/CdS solar cells (see *Tellurides* in the following).

***Tellurides*:** This group of II–VI compounds is represented by zinc telluride (ZnTe) and cadmium telluride (CdTe). The former features energy gap $E_g = 2.2$ eV and because it lends itself to technological manipulations including doping and adjustments of lattice constants, is among important in practice II–VI semiconductor compounds. CdTe ($E_g = 1.5$ eV) is used to manufacture solar cells. Combination of CdTe with CdS expands significantly portion of the solar spectrum captured by such cell. Similarly to CdSe, also CdTe is used in the form of zero-D nanodots which when reduced in size to single nanometers in diameter expand emission spectrum of CdTe toward shorter wavelengths by decreasing width of the energy gap.

Ternary and quaternary II–VI semiconductor compounds: The reason for alloying various II–VI binary compounds into ternary and quaternary compounds is the same as in the case of discussed earlier ternary III–V compounds. By mixing and matching various binary II–VI alloys, continuous, independent modifications of the energy gap E_g and the lattice constant a of the resulting compound can be achieved. For instance, by alloying CdTe and ZnTe into $Cd_xZn_{(1-x)}Te$, or CZT, and changing its composition by changing x, the bandgap of the ternary alloy can be varied from 1.5 eV for $x = 1$ to 2.2 eV for $x = 0$. The CZT is a crystal used in ultra-short wavelength radiation detection.

Even finer tuning of the bandgap characteristics within similar range, but at the expense of more complex processing resulting from the need to control all four elements of an alloy, can be accomplished by mixing two binary compounds ZnSe and CdTe into quaternary compound $Zn_{1-y}Cd_ySe_{1-x}Te_x$.

Another example of II–VI ternary compound is concerned with mercury cadmium telluride, HgCdTe, also known as MCT, which is a semiconductor compound of significant technological importance. The MCT is an alloy of mercury telluride HgTe which is a semi-metal with no energy gap, and CdTe which is semiconductor featuring $E_g = 1.5$ eV. By controlling composition of the ternary alloy, its energy gap can be varied as desired from 0 eV and 1.5 eV.

2.4.3 *Geometrically Confined Semiconductor Materials*

Following on the discussion of the properties of nanoscale materials in section 1.5 and concerned with reduction of materials geometry introduced in Fig. 1.15, purpose of the brief remarks below is to provide additional information regarding solid-state nanoscaling in the context of specific inorganic semiconductor materials.

2D semiconductor materials: Besides 2D carbon known as an introduced earlier graphene, also one-atom thick silicon known as silicene has been experimentally obtained and the presence on the energy gap in its energy band structure was confirmed. Comprised of hexagonally bonded silicon atoms, silicene, is particularly promising because of its compatibility with silicon process technology. If synthesized on the insulating substrate, silicene can be a viable contributor to the advancements in semiconductor electronics.

In addition to carbon (*graphene*) and silicon (*silicene*), two remaining elements in the group IV of the semiconductor Periodic Table (Fig. 2.10), namely germanium and tin, can be also obtained in 2D configurations called *germanene* and *stanene*, respectively. More detailed understanding of the fundamental properties of these 2D materials is needed to determine their potential in the specific device applications (Bonca and Kruchinin, 2010).

Furthermore, there are other 2D materials offering potentially a range of applications in semiconductor devices performing electronic, photonic and biomedical functions because, unlike graphene, some of them feature energy gap. In addition, in contrast to elemental 2D materials, their properties can be tuned by adjusting their composition, structure, and size. Among them, 2D compounds known as dichalcogenides which include two atoms of a chalcogen (sulfur, tellurium or selenium) and transition metal (molybdenum, tungsten or titanium) and referred to as transition metal dichalcogenides (TMD) are of particular interest often form atomic layered structures with one layer of metal sandwiched between two layers of chalcogen atoms.

1D materials: The 1D nanostructures (Fig. 1.15) are represented by nanotubes, most commonly carbon nanotubes referred to as CNTs, and nanowires in which case silicon nanowires referred to as SiNWs are the best developed.

Carbon nanotubes display unique electrical, mechanical, and thermal properties. As shown in Fig. 2.11 the CNT can be seen as a sheet of graphene rolled up into cylinder few nanometers in diameter which under the right conditions can reach the length in excess of 10 centimeters. The nanotubes can be single-walled (SWNT), double-walled (DWNT), and multi-walled (MWNT) depending on the number of concentrically rolled-up graphene sheets.

CNTs can display semiconductor or metallic properties depending on their structure resulting from the direction graphene is rolled up into a nanotube and producing either so-called *zigzag* CNT or *armchair* CNT.

Due to their very low resistivity and ability to carry very high density current, the main applications of CNTs in semiconductor device engineering at this point are in the interconnect technology in ultra-small geometry integrated circuits discussed in Chapter 4.

0D materials: Known as nanodots or quantum dots, 0D materials represent yet another class of the nanoscale inorganic semiconductors. Nanodots can be as small as few nanometers in diameter and can be synthesized using metals, insulators, and semiconductors including available commercially silicon nanodots. In the case of semiconductors, of particular interest are nanocrystalline quantum dots (NQD) physical properties of which, most notably width of the energy gap and consequently wavelength of emitted light, can be tuned by changing diameter of the dot (Price *et al.*, 2009).

Good example of this behavior is provided by the introduced earlier cadmium selenide, CdSe, which in the standard 3D form at the room temperature features energy gap E_g = 1.75 eV and can be used for emission and detection of infrared radiation. The nanoconfinement of CdSe to nanodot geometry increases width of the energy gap, and consequently, shortens the wavelength of emitted light. When reduced to the geometry of 6.0 nm in diameter, the energy gap of CdSe increases to 2.0 eV, and the wavelength of the emitted light shifts from the near-infrared 700 nm for 3D CdSe, to 610 nm (red light). Further reduction of the CdSe quantum dot diameter to 2.0 nm, increases its energy gap to 2.75 eV, and shortens wavelength of the emitted light to 450 nm (blue light).

The size dependent modifications of the physical characteristics (energy gap in particular) of semiconductor nanodots is a foundation upon which applications of NQDs in the light emitting and light detecting devices are based (Tan *et al.*, 2008).

2.5 Organic Semiconductor Materials

In addition to the broad range of inorganic semiconductors discussed in this chapter, selected organic materials (i.e. materials consisting primarily of carbon and hydrogen) display semiconductor properties, and in addition, other distinct features allowing for a range of unique applications (Stalliga, 2009).

Organic materials are in general weak conductors of electricity. Some of them, however, upon injection of charge carriers (electrons or holes), or after being properly chemically engineered, take on characteristics of semiconductors by allowing control of charge distribution using external electric field and displaying ability to emit and detect light. As a result, this class of organic compounds is being appropriately referred to as organic semiconductors (Sabba, 2017; Alcacer, 2018).

The progress in the technology of organic semiconductor materials happens not because they feature superior to inorganic semiconductors characteristics, but because they allow expansion of applications of semiconductors into the areas in which conventional thin-film inorganic semiconductors cannot play any useful role. In this context, the fact that in terms of tangible measures, such as, for instance, charge carrier mobility, organic semiconductors are inferior to their inorganic counterparts is not a deciding consideration.

The following factors distinguish organic, or "plastic", semiconductors from their inorganic counterparts: (i) organic semiconductors maintain their fundamental physical properties even if drastically bent, and thus, are compatible with flexible substrates, (ii) organic semiconductors are used only as the thin-film, non-crystalline materials, (iii) organic semiconductors are transparent to visible light, and (iv) organic semiconductors are low-cost materials which can be processed into functional devices using relatively simple manufacturing technology. These characteristics open up possibility of formation of functional semi-transparent semiconductor devices and circuits on the flexible substrates.

Organic semiconductors are based on either small molecules (monomers) shown in Fig. 2.14(a) or polymers comprised of small molecules shaped into chains (Fig. 2.14(b)). In both cases weak van der Waals forces are responsible for the cohesion of resulting plastic-like solids. In terms of chemical composition, the most commonly used representative of small-molecule organic semiconductors is pentacene, $C_{22}H_{14}$. The common polymeric organic semiconductors are conjugated polymers, i.e. polymers composed of two linked compounds chemistry of which is beyond the scope of this discussion. Both single-molecule and polymer semiconductors are readily available commercially.

The way organic semiconductors are put together is responsible for the fundamentally different electric charge transport mechanisms in inorganic and organic semiconductors. In the former case electrons are moving as

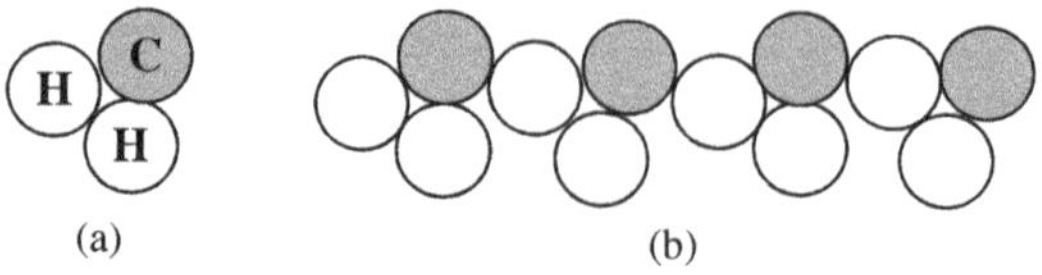

Fig. 2.14. (a) Small molecule (monomer) organic semiconductor and (b) polymer organic semiconductor.

delocalized plane waves and as such are subjected to limited scattering, and hence, feature relatively high mobility such as, for instance, 1500 cm^2/Vs for Si at room temperature (Table 2.1). In the case of organic semiconductors, the charge transport is based on carriers hopping between localized states associated with organic molecules. In the process, electrons are subject to significant scattering which results in a very low electron mobility in organic semiconductors typically in the range of 1–3 cm^2/Vs with pentacene distinguishing itself among organic semiconductors by the highest carrier mobility.

Despite inferior to inorganic semiconductors electronic properties, organic semiconductors are broadly used in the range of applications with which common inorganic semiconductors are simply not compatible. These include transparent, flexible (Wong and Salleo, 2009), and printable electronic and photonic devices and circuits which will be considered further in Chapter 3.

2.6 Semiconductor Materials Selection Criteria

As it was pointed out earlier, there is a strong correlation between the properties of semiconductor material, and performance of devices designed to perform specific functions which are fabricated using it. What it means is that some semiconductor materials by virtue of their inherent properties are suitable for carrying out some device functions, while are not suitable to carry out some other functions. For instance, semiconductor material featuring high electron mobility, but narrow energy gap is suitable for high-speed device applications, but may not suitable for high-power/high-temperature operation. On the other hand, wide-bandgap semiconductors featuring low electron mobility are very well suited for the latter applications, but not for the former. Furthermore, type of the bandgap, direct or indirect will predetermine usefulness of any given semiconductor in light emitting devices. Semiconductors with indirect bandgap, in which radiative recombination is inefficient, will not be considered for light emission device applications. Carrier mobility is another important key criterion for selection of semiconductor materials for specific device applications in which speed of operation is of interest.

Table 2.3 considers device performance implications of the physical characteristics of semiconductor materials introduced in this chapter. The left side column in the table identifies selected physical characteristics of semiconductor materials while the column on the right shows how any given characteristic may impact performance of semiconductor devices.

Table 2.3. Device implications of select physical characteristics of inorganic semiconductor materials.

Material Characteristics	Device Implications
Energy gap width, E_g	*Electronic devices* – wide bandgap (E_g >2 eV) desired for better power, temperature handling
	Photonic devices – defines light absorption and emission characteristics
Type of energy gap	*Indirect* – inefficient radiative recombination
	Direct – efficient radiative recombination (suitable for light emitting devices)
Electron mobility	Determines speed of device operation
Saturation electron velocity	Determines speed of device operation at high internal electric field
Oxidation characteristics	Determine ability to form high-quality native oxide
Susceptibility to doping	Both *n*-type and *p*-type doping should be possible
Defect density	Defects deteriorate device performance
Breakdown strengths	Determines resistance to high electric field
Thermal conductivity	Determine ability to dissipate heat generated during device operation
Thermal stability/ melting point	Determines resistance to high temperature – important during device fabrication and elevated temperature during device operation
Radiation hardness	Determines degree of undesired sensitivity of semiconductor material to high energy radiation
Mechanical stability	Needed to prevent damage of the semiconductor substrate during device fabrication

Alternative criterion defining usefulness of materials in various applications can be based on the mechanical characteristics and size of the substrate upon which semiconductor devices and circuits are formed. While general guidelines defined in Table 2.3 remain valid, somewhat modified consideration are applicable to the wafer-based, thus rigid, electronic and photonic devices formed on limited-area substrates, and to flexible and large-area electronic and photonic devices. What it means is that the physical properties of any given semiconductor material and the type of the substrate used are not the only criteria upon which selection of materials is based. In some cases, availability of material in desired crystallographic structure, its cost, as well as broadly understood manufacturability related characteristics will influence the material selection process.

To summarize the discussion of semiconductor materials, Fig. 2.15 illustrates how selected elemental and compound inorganic semiconductors, and their energy gap E_g, type of energy gap, and electron mobility are aligned with respect to the electromagnetic spectrum.

2.7 Materials Comprising Semiconductor Device

In previous sections in this chapter, semiconductor materials were considered in terms of their inherent properties with little reference to device applications. In sections below materials technology is considered with general device applications in mind. This includes recognition of the fact that not only semiconductors but also metals (conductors) and insulators (dielectrics or ferroelectrics) are needed to ensure expected operation of any given semiconductor device.

Looking at common semiconductor device such as metal–oxide–semiconductor field effect transistor (MOSFET, to be discussed in Chapter 3), Fig. 2.16 shows the same cross-section as in Fig. 2.5 with materials it is comprised of identified.

As the above figure shows, the device is formed on the substrate which, as it was discussed earlier, most often is a single-crystal semiconductor but can also be an insulator featuring divers compositions and mechanical properties. Selection of the substrate material is based on the function given device is designed to perform. Discussion in this book includes ample examples of the role substrate technology plays in semiconductor device engineering.

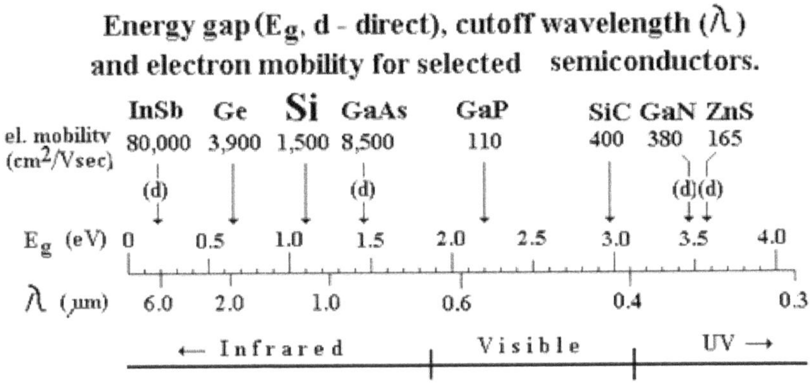

Fig. 2.15. Energy bandgap, type of energy gap, cut-off wavelengths and electron mobility for various semiconductor materials.

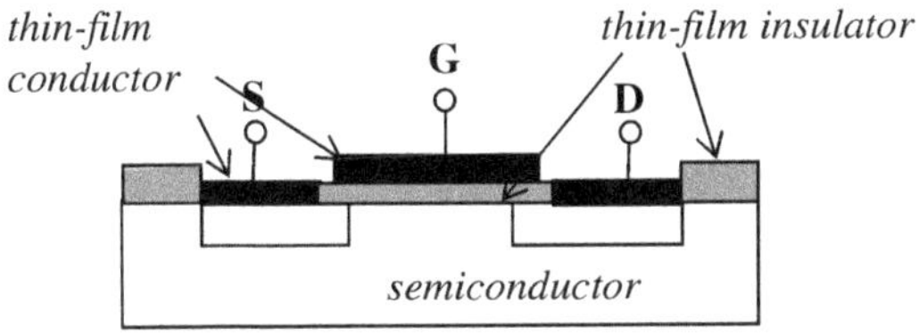

Fig. 2.16. Cross-section of the common semiconductor device (MOSFET, see the following chapter) identifying its parts made out of conductors, semiconductors and insulators.

Other than the substrate, schematics in Fig. 2.16 show thin film conductors as well as insulators acting as the integral parts of the device. The former serve the purpose of contacts and interconnects while the latter fulfill multiplicity of functions depending on the design and purposes of semiconductor device, they are the part of.

Furthermore, there are various other materials which are not part of the final device, but involved in the manufacturing processes. Photosensitive materials known as photoresists are the prime example of materials in this class.

What the comments above underscore is that semiconductor device engineering is a heavily interdisciplinary technical domain in which materials science and engineering play crucial role.

2.7.1 *Bulk Single-Crystal Formation*

At the core of semiconductor device engineering are single-crystal semiconductors, and hence, the methods used to form single-crystal semiconductors are of vital interest in semiconductor device technology. The reason is a very strong correlation between the quality of the crystal (see defects in Fig. 2.4) and performance of any device based on crystalline semiconductor material. In this section, methods used to grow bulk single-crystal semiconductor materials are reviewed. In the following section, formation of thin-film single-crystal material through the process of epitaxial deposition is discussed.

CZ single-crystal growth: The most common in the manufacture of bulk single-crystal semiconductors (as opposed to thin-film single crystals) is the method of Czochralski crystal growth (Fornari, 2018) referred to in short as a CZ method. The CZ method is best explained using growth of single-crystal silicon as an example. In this technique, nuggets of ultra-pure polycrystalline Si are molten in the crucible with high purity quartz or silicon carbide liner. To establish *n*-type or *p*-type conductivity and desired doping level, dopants in precisely controlled amounts are added to the melt. Subsequently, properly oriented with respect to the surface of the melt piece of high-quality single-crystal Si, referred to as crystal seed, is immersed in the melt (Fig. 2.17(a)) and then very slowly pulled out of the melt with rotation (Fig. 2.17(b)). During the pulling process, crystallization of molten Si is taking place at the interface between molten silicon and the single-crystal ingot, also referred to as a single-crystal rod. The crystallographic structure of the single-crystal ingot pulled out of the melt (Fig. 2.17(c)) is the same as the crystallographic structure of the seed.

The nature of the single-crystal pulling process in CZ method predetermines circular shape of the ingot. Diameter of the ingot defines in turn diameter of the single-crystal wafers that will be cut out of the ingot and in the case of silicon can be as large as 450 mm.

Among single-crystal formation methods the CZ method is the one that allows (i) the largest in diameter ingots, and thus, the largest wafers, (ii) single-crystals featuring the lowest density of crystallographic defects, and (iii) the most uniform radial distribution of dopant atoms along the diameter of the ingot. Due to these advantageous characteristics, the CZ crystal growth is the most broadly used bulk single-crystal fabrication method in semiconductor technology.

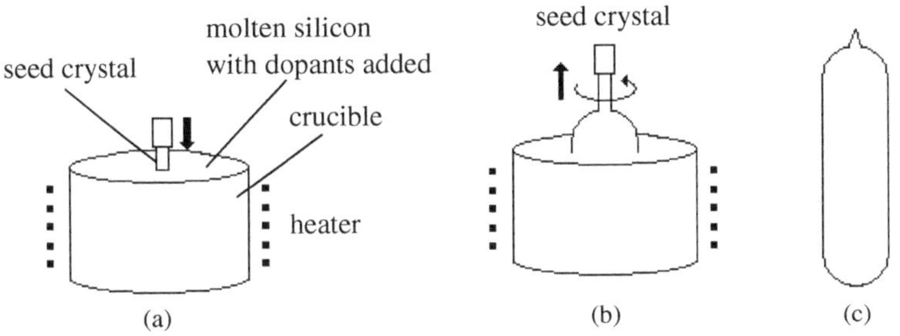

Fig. 2.17. Schematic representation of the Czochralski (CZ) single-crystal growth process, (a) seed crystal is immersed in the liquid silicon, (b) single-crystal silicon is pulled out of the liquid, and (c) the result of the process in the form of the single-crystal silicon ingot.

There are other methods of bulk single-crystal semiconductor formation available which are used where CZ method cannot produce single-crystal material featuring specific characteristics in terms, for instance, of resistivity or cannot be used because of the inherent properties of a given semiconductor material being incompatible with the CZ growth conditions (Campbell, 2020).

The first limitations come to play where crystal to be formed needs to be free from any contamination with alien elements in order to feature required in some applications very high resistivity of single-crystal semiconductor material such as silicon. The CZ growth methods by virtue of the molten material remaining in contact with crucible at very high temperature, cannot prevent contamination of silicon crystal with, for instance, oxygen or carbon leaching out of the crucible. In this case, alternative method known as float-zone (FZ) crystallization must be used to form single crystal.

The second listed above limitation of the CZ growth process results from the fact that some semiconductor materials are not compatible with high temperature of the CZ process. It concerns in particular semiconductor compounds in which vapor pressure of constituting elements varies significantly. In the case of gallium arsenide, GaAs, for instance, vapor pressure of As is significantly higher than the vapor pressure of Ga. In such case, maintaining equilibrium composition of the compound in the melt in the course of CZ process is difficult due to the excessive evaporation of one element (arsenic, As, in the case of GaAs). The Bridgman method of single-crystal growth which was conceived to work around these limitations is used to form single crystals of some key compound semiconductors (Rockett, 2008).

The CZ method has also limited use in single-crystal formation of semiconductors featuring very high melting point such as, for instance, silicon carbide, SiC. With melting point of 2,730 °C, SiC requires a method of seeded sublimation to form single crystal. In this method SiC powder is heated to 2,200 °C at the reduced pressure under which conditions SiC sublimes i.e. transitions from the solid phase directly to the vapor phase. The SiC vapor reaches positioned nearby SiC single-crystal seed where it coalesces forming a boule of single-crystal SiC.

Somewhat special in terms of single-crystal growth is the case of gallium nitride, GaN. This indispensable in the range of photonic and high-power electronic applications semiconductor material cannot be obtained in the bulk single-crystal form using any of the method listed above. To obtain bulk single-crystal GaN in the shape and size compatible with the demands of commercial device manufacturing, extreme measures in terms of pressure and temperature, such as in the case of ammonothermal autoclave, must be taken to form single-crystal GaN.

The cost of the process as well technical challenges preventing formation of the large single-crystal GaN wafers call for the alternative solutions that are being implemented to ensure progress in GaN device technology (see Section 2.7 for more detailed considerations of this topic).

2.7.2 *Thin-Film Single-Crystal Formation*

In addition to bulk single-crystal growth, methods forming thin-film single-crystal semiconductor materials are among cornerstone processes in semiconductor engineering. In this context the concept of epitaxial deposition is of fundamental importance.

Epitaxial deposition: The term *epitaxial* deposition refers to the process known as epitaxy (from Greek "ordered deposition") which forms a layer of the solid on the single-crystal substrate in such way that the crystallographic structure of deposited material reproduces exactly crystallographic structure of the substrate (Fig. 2.18(a)). By definition then, a layer formed by means of epitaxial deposition, or epitaxial layer (epi layer in short) is a single-crystal material featuring the same lattice constant a_f as the lattice constant of the substrate a_s.

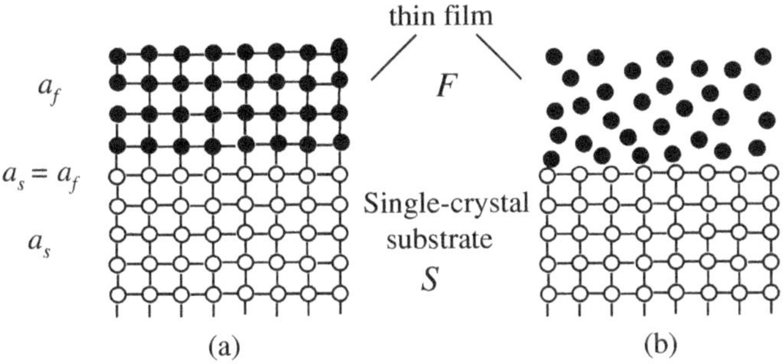

Fig. 2.18. (a) Epitaxial deposition and (b) non-epitaxial deposition.

Depending on the deposition conditions, the same in terms of chemical composition material can be deposited on the substrate in such way that the former will not reproduce crystallographic structure of the single-crystal substrate, and deposition of the polycrystalline or amorphous film will result (Fig. 2.18 (b)). If this is the case, then the process obviously does not involve an epitaxial deposition, but rather conventional deposition in which crystallinity of deposited film is not a key consideration.

The chemical composition of the material deposited through epitaxy can be the same as a chemical composition of the substrate (homoepitaxy), or different (heteroepitaxy), as long as the lattice constants a (see Fig. 2.2) of both are matched (lattice matching). As can be concluded from this discussion, the process of epitaxial deposition is by default involving a single-crystal substrate of which the surface must be pristinely clean to allow undisturbed epitaxial growth. If the surface of the substrate is covered with any non-crystalline residual film, even ultra-thin, the atoms which are meant to form an epi layer will be shielded from the substrate, and hence, won't be able to reproduce crystallographic structure of the substrate. Furthermore, in order to allow adsorbed atoms to align themselves with the atoms in the substrate, and then to form bonds with the atoms on the surface using the least amount of energy, epitaxial deposition processes need to be carried out at elevated temperature.

Thin layers of single-crystal epitaxially deposited semiconductor feature a range of characteristics which cannot be obtained in any other way, and which are amply exploited in semiconductor device engineering. The key features making epitaxy unique are as follows.

First, defining characteristic of epitaxial deposition is that the crystal structure of the deposited film reproduces exactly crystal structure of the substrate upon which film is formed. Second, dopant atoms are introduced into the growing film during the epitaxial growth process such that conductivity type (*n*- or *p*-type), as well as a doping level of the epitaxial layer, can be set independently of the doping of the substrate upon which the epitaxial film is deposited. Third, since dopants introduction take place during the growth of the epitaxial film, dopant atoms are uniformly distributed across the thickness of the film.

Another characteristic of epitaxial deposition is that with some deposition techniques, epitaxial growth can be controlled with precision allowing growth of the films in the thickness range of single nanometers. Combined with mentioned above ability to establish conductivity type and doping level independently of the substrate, epitaxial deposition is a tool allowing discussed earlier bandgap engineering in the semiconductor material systems.

Furthermore, neither gases from which epi film is formed nor the film itself are ever in contact with alien solids such as crucibles involved in the bulk crystal growth. As a result, epitaxial layers are typically chemically purer than the bulk single-crystals upon which they are deposited.

At the downside of epitaxial deposition technique is that the structural defects of the surface of the substrate upon which epitaxial layer is formed are reproduced and augmented in the growing film. Also, epitaxial deposition is a process requiring elevated temperature. Depending on how it is implemented (see Chapter 5 for more detailed discussion), and what materials are deposited, temperature of epitaxial deposition may vary from 500 °C to 1,100 °C.

Lattice matched and lattice mismatched epitaxial deposition: With the features outlined above epitaxial deposition is a process of key importance in the manufacture of the broad range of advanced semiconductor devices. Based on the purpose epitaxial deposition serves and the way it is implemented, lattice matched and lattice mismatched epitaxial processes are distinguished. Figure 2.19(a) illustrates the former process in which lattice constants *a* of the substrate and epi film are exactly the same, and hence, both feature the same chemical composition. In such case of homoepitaxy, silicon epi layer can be deposited on single-crystal silicon substrate, for instance, for the purpose of forming a uniformly doped film featuring different conductivity type (*n*- or *p*-type) than the substrate.

Lattice mismatched epitaxy is implemented based on different principles and serves different than lattice matched homoepitaxy purposes. In general, crystals featuring different chemical composition, and thus, different lattice constants, cannot be integrated into one material system by means of epitaxial deposition without formation of the defects in the growing crystal. However, there are situations in which controlled lattice mismatch between substrate and epi layer can not only be overcome, but is desired. This is because lattice mismatch introduces strain in the lattice which, as indicated in Chapter 1, results in the increased electron mobility in the strained material which in turns leads to the faster operation of electronic device featuring strained crystal lattice in its parts through which electrons are moving.

Figure 2.19(b) illustrates principles of the strained layer heteroepitaxy. Consider single-crystal substrate *S* featuring lattice constant a_s upon which film *F* featuring different lattice constant a_f is deposited. As a film grows, strain resulting from the relative displacement of atoms in the lattice (lattice mismatch) continues to build up until at a certain film thickness the strain energy is relieved through rearrangement of bonds alignment in the lattice resulting in the formation of defects, mainly in the form of dislocations, in the crystal lattice (Fig. 2.19(a)). At this point lattice constant a_f of relaxed lattice of the film will return to its unstrained value. However, if the film growth is stopped below critical thickness h_c, i.e. in the thickness regime in which lattice mismatch can be accommodated by strain, a highly strained single-crystal film *F* called a pseudomorphic film will be formed on the substrate *S* (Fig. 2.19(b)). Critical thickness h_c decreases as the lattice mismatch $f = a_{f-}\,a_s/a_{verage}$ increases. In practical strained layer heterostructures, the lattice mismatch typically

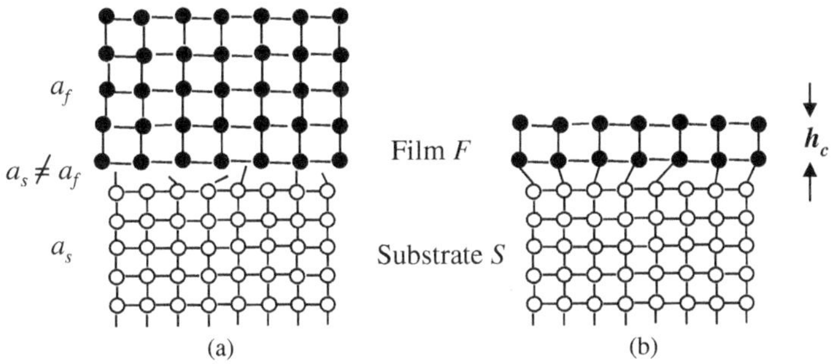

Fig. 2.19. (a) When the lattice constants *a* of the substrate and deposited film do not match, defects are formed at the interface, (b) strain in the mismatched lattices can be accommodated providing thickness of the deposited film does not exceed critical value h_c.

does not exceed 5% and the thickness of the strained film ranges from 1 nm to 20 nm. In terms of specific materials, epitaxial deposition of pseudomorphic silicon germanium, SiGe, on silicon exemplifies strained layer hetroepitaxy.

In the case two lattice mismatched materials need to be combined into a single material system without introducing strain in the lattice, a buffer layer, or layers, allowing for a gradual transition from one lattice configuration (lattice constant) to another without straining of the crystal lattice can be employed. A term graded layer is used in this context to underscore gradual transition between different lattice constants materials by manipulation of their chemical composition. A combination of heteroepitaxy, strained layer heteroepitaxy with graded layer technology allows significant flexibility in fabrication of complex materials systems serving a broad range of electronic and photonic functions. An example of such material system is a superlattice comprised of several layers of ultra-thin films (typically 1 nm–2 nm thick) which can be lattice matched, and hence, strain-free, or lattice mismatched, and hence, forming a strained layer superlattice (SLS).

The case of superlattice is exemplified by the GaAs/AlAs material system. The GaAs/AlAs material system is unique in that it features lattice matching for all Al_xGa_{1-x} As alloy compositions. As discussed earlier, with x varying from 0 to 1 the bandgap of this ternary compound changes from $E_g = 1.42$ eV to $E_g = 2.16$ eV accomplishing bandgap engineering without introducing strain in the lattice. The latter can be illustrated using GaAs/InAs material system as an example. These two III–V compounds, GaAs and InAs, cannot be lattice matched at any composition, and hence, are well suited to form strained layer superlattice.

In either configuration, discussed earlier quantum wells can be formed in the layers of narrower bandgap material sandwiched between wider bandgap materials. In the cases considered here, it would be an ultra-thin AlGaAs sandwiched between layers of GaAs, and an ultra-thin InGaAs sandwiched between layers of GaAs in the strained layer superlattice.

2.8 Substrates

A key element in semiconductor engineering is a mechanically coherent piece of the solid with which referred to as a substrate and with which process of making device starts. Substrate can be (i) a bulk semiconductor material which properties are locally altered to build into it functional electronic or photonic devices or (ii) a bulk semiconductor material or insulator upon which thin-film semiconductor system designed to perform electronic or photonic functions is constructed. In the former case substrate participates in device operation. In the latter case, substrate is used as a mechanical support for the multilayer thin-film material system built upon it, but which at the same time features desired electrical and optical characteristics. Most commonly, substrates used to manufacture semiconductor devices are shaped as rigid circular slices referred to as wafers made of single-crystal semiconductor material obtained as shown in Fig. 2.20. In special applications, device technology departs from this standard by using variety of non-circular substrates made out of various materials either rigid or flexible.

In the following review, various types of substrates used in semiconductor device engineering are considered distinguishing between semiconductor and non-semiconductor substrates.

2.8.1 *Semiconductor Substrates*

As pointed out earlier, in the majority of both electronic and photonic applications term "substrate" is synonymous in semiconductor terminology with a circular-shaped wafer of single-crystal semiconductor. Depending on material and process needs, single-crystal semiconductor wafers vary in diameter from less than 20 mm to 450 mm with thickness varying from less than about 0.1 mm to about 1.0 mm depending on the size of the wafer and its use. In general, with wafers getting larger in

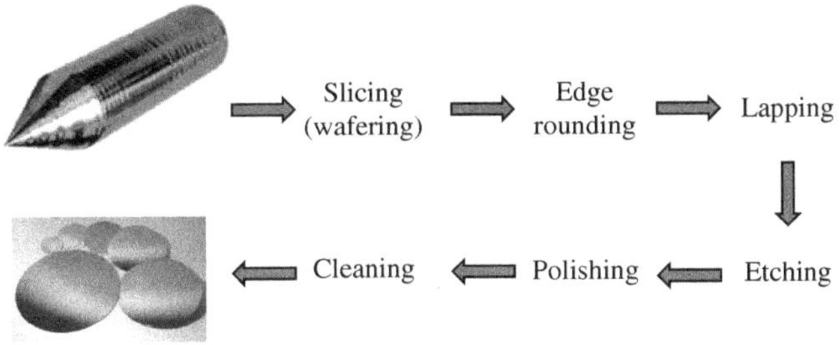

Fig. 2.20. From the single-crystal bulk ingot to the wafers used to fabricate devices.

diameter, increased number of devices can be formed on the wafer as a result of which cost of the electronic functions performed by the individual device formed on the wafer is lower.

Bulk wafers: Bulk single-crystal material is typically obtained in the form of an ingot using methods discussed in earlier. Key operations involved in the conversion of single-crystal ingots into production-ready wafers are shown in Fig. 2.20. First, an ingot is machined to assure uniformity of diameter along its length. Then, in the course of the process called wafering, single-crystal ingots are sliced into wafers typically using high precision multi-wire saw configured to cut multiple wafers at a time from the same crystal. Ingots are sliced into wafers along designated crystallographic planes to establish desired surface orientation, for instance, (100) or (111). This element of the process is important because the arrangement of atoms on the surface defined by surface orientation plays a role in defining characteristics of devices formed on such surfaces.

Subsequently, edge rounding, lapping, etching and polishing operations are performed on each wafer to establish mirror-like surface quality typically on one side of the wafer (sometimes wafers are polished on both sides). From the device manufacturing perspective, final polishing step is of special interest because it determines quality of the wafer's surface and near-surface region.

A special role in the wafer fabrication sequence (Fig. 2.20) is played by the final cleaning operation. Assisted by mechanical scrubbing using soft brushes needed to remove polishing slurry, this multistep process is designed to remove all chemical contaminants as well as the tiniest particles from the wafer surface (see more detailed discussion of surface contaminants in Chapter 4 and cleaning processes in semiconductor device manufacturing in Chapter 5).

The discussed above general principle underlying fabrication of substrate wafers applies also to square-shaped wafers such as those used in the manufacture of solar cells. The difference is in (i) the geometry of the starting ingot which needs to be square or rectangular in cross-section, (ii) crystal structure of the ingot which can be either single crystal or polycrystalline, and (iii) the role of the polishing step which is needed to define geometry of the wafer, but is not meant to establish mirror-line surface finish which would promote undesired reflection of the sunlight from the surface in the working solar cell. Just the opposite, surfaces of the solar cells are textured on purpose to prevent reflection.

Barring few exceptions, most of inorganic semiconductor materials considered earlier in this chapter can be processed into a substrate wafers following procedures outlined in general terms above. Depending on material, diameter of the wafers and their crystalline quality vary significantly with silicon lending itself to the manufacture of the largest and the highest quality wafers. In order to build into them desired features such wafers are commonly subjected to additional treatments which are briefly reviewed below using Si wafers as an example.

Engineered silicon wafers: Processes illustrated in Fig. 2.20 produce high-quality bulk wafers (Fig. 2.21(a)) meeting requirements of the typical semiconductor device manufacturing process. In the high-end device manufacturing, however, such homogeneous Si wafers, commonly referred to as bulk wafers, need to be engineered further to meet specific device related requirements. Directions wafer engineering may follow are illustrated in Fig. 2.21 and are considered in the following discussion.

***Denuded zone formation*:** Term *denuded zone* refers to the very thin part of the wafer immediately adjacent to its top surface from which some excessive structural defects and/or alien elements (contaminants) are displaced into the bulk portion of the wafer by means of the gettering processes (Fig. 2.21(b)). In semiconductor terminology term *gettering* means enforcing motion of contaminants and/or structural defects in

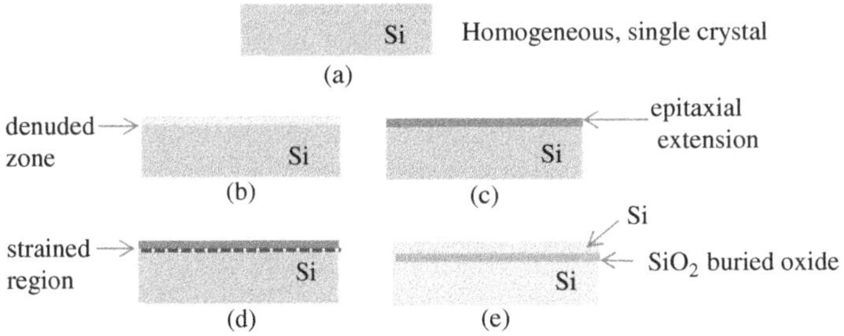

Fig. 2.21. Illustration of the ways homogeneous, single-crystal Si wafer can be engineered to meet the needs of various device applications.

semiconductor crystal lattice away from the top surface of the wafer into its bulk and trapping them there. In this way, a free from defects and contaminants denuded zone immediately adjacent to the top surface of the wafer, i.e. region into which active devices will be built-in (see discussion in Chapter 3), is created.

Typically, available on the market high-quality CZ silicon wafers feature denuded zone formed with help of the gettering. Gettering processes are implemented either solely via series of thermal treatments to which wafer is subjected in strictly pre-determined sequence (intrinsic gettering), or by using external interactions to alter the distribution of stress in the wafer which in combination with thermal treatment enforces motion of certain types of defects and certain contaminants away from the top surface (extrinsic gettering).

***Epitaxial extension*:** As discussed earlier, process of epitaxial deposition allows formation of the very-thin layer of single-crystal material in such way that the crystallographic structure of the deposited film exactly reproduces crystallographic structure of the substrate. At the same time, type of doping, *p* or *n*, can be the same, or different than the substrate. Another advantage of an epitaxial extension (Fig. 2.21(c)) is that the chemical purity of the epi-layer and physical characteristics of its surface (smoothness) are superior to that of the substrate and its surface which was subjected to various mechanical treatments (lapping, polishing) during wafer manufacturing (Fig. 2.20). Also, heteroepitaxial deposition is possible as an epitaxial extension as long as there is a lattice matching between substrate and deposited film or strained layer at the interface is desired. When substrates comprised of thicker lattice mismatched crystals are needed, a technique of wafer bonding discussed later in this section may be employed.

With advantages listed above, epitaxial extension is a common modification of the bulk wafers used in advanced semiconductor device manufacturing. Depending on application, thickness of epitaxial extension may vary from as thin as few nanometers to as thick as tens of micrometers.

***Strained layer*:** Strained-layer heteroepitaxy discussed earlier and illustrated in Fig. 2.19 is yet another technique that is being used to engineer substrate wafer toward building into it desired characteristics which in this case is a strained top surface layer (Fig. 2.21(d)).

Silicon-on-Insulator (SOI): Acronym SOI in common usage stands for *Silicon-on-Insulator*, but in the broader sense can also be used in reference to *Semiconductor-on-Insulator* substrate wafers. Figure 2.21(e) shows SOI wafer which represents yet another way in which silicon wafer can be engineered (Arora *et al.*, 2016). In this case, the goal is to form underneath its near-surface region a thin layer of an insulator which in the case of Si wafers is its native oxide, silicon dioxide (SiO_2). Since the oxide can be envisioned as being "buried" in silicon, a term buried oxide, or BOX in short, is commonly used in reference to such oxide. A layer of silicon on top of the BOX is an active silicon layer in which functional devices are formed, while the part of the wafer underneath BOX is acting merely as a mechanical support.

There are two different approaches to the fabrication of SOI wafers with buried oxide (Fig. 2.22). First is based on the discussed later in this section technology of wafer bonding. As Fig. 2.22(a) illustrates, two bulk Si wafers A and B, this time both with oxide covered surfaces to be bonded, and one (wafer A) implanted with hydrogen, are pressed against each other at elevated temperature to establish a permanent bond with oxide in the bonding plane forming a buried oxide of the future SOI substrate (Fig. 2.22(b)). Next, the wafer A is separated by thermally promoted crystal cleaving along the stressed by hydrogen implantation plane (Fig. 2.22(c)). The process known as SmartCut™ allows splitting of the

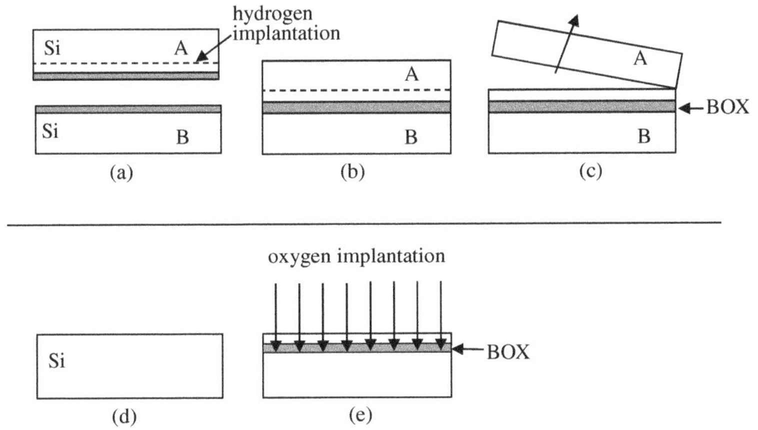

Fig. 2.22. Simplified illustration of the SOI wafer fabrication by mean of (a)–(c) SmartCut™ process (SOITEC, SA) and (d) and (e) SIMOX process.

wafer A after bonding without damaging it, and hence, allows wafer A to be reused.

An alternative to bonding commercial method of SOI substrate fabrication is based on the technique of ion implantation discussed in more details in Chapter 5 of this volume (Fig. 2.22(d, e)). In this case oxygen ions are accelerated toward bulk Si wafer, impinge on its surface, and then penetrate near-surface region of the wafer to the depth determined by their energy resulting from acceleration. Following implantation, wafer is subjected to thermal treatment needed to remove damage to the Si crystal inflicted by high-energy oxygen ions and at the same time to enforce chemical reaction between implanted oxygen and silicon to form a buried layer of silicon dioxide, SiO_2 (buried oxide (BOX) Fig. 2.22(d)). The SOI fabrication method based on oxygen implantation is known as Separation by IMplantation of OXygen (SIMOX) process.

Yet another variation of the SOI wafer technology involves Silicon-on-Sapphire (SOS) considered briefly later in this section.

Just like in the case of other types of engineered semiconductor wafers, SOI wafers are fabricated by specialized industrial wafer manufacturers and are commercially available to the devices manufacturers in configurations and sizes conducive with their respective needs.

Wafer bonding: The process which permanently bonds (fuses) two wafers into a single mechanically coherent substrate without using adhesives. This versatile technique allows formation of semiconductor substrates which are difficult to obtain using other methods. The problem is that not all semiconductors can be obtained as single crystals in the size and shape compatible with wafering. With wafer bonding, pieces of such material can be bonded to larger wafer which will provide mechanical support and facilitate wafer handling during device fabrication processes.

In the process of wafer bonding two wafers made out the same material (homobonding), for instance, two silicon wafers, or two single-crystal wafers of different materials (heterobonding) featuring mismatched crystal lattices, for instance, GaN and Si, can be bonded into a single substrate.

Figure 2.23 illustrates the process of wafer bonding. Surfaces to be bonded must be very smooth and clean, hence, very thorough cleaning of both wafers is an integral part of the bonding procedure. Following surface preparation, pressure is applied to wafers A and B shown in

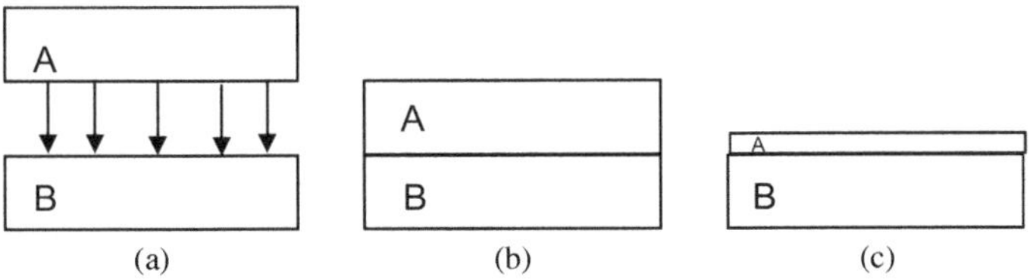

Fig. 2.23. Wafer bonding process, (a) two wafers are pressed against each other at an elevated temperature and (b) are permanently bonded, (c) top wafer is thinned to the desired thickness.

Fig. 2.23 at the elevated temperature using designated wafer bonding equipment typically operating under vacuum conditions. The nature of the forces responsible for the permanent bonding of two wafers depends on several factors including chemical makeup of the bonded materials, as well as conditions of the bonding process. The most common types of interactions between two solids brought to physical contact include van der Waals forces, electrostatic (Coulombic forces), and capillary forces, often in combination (Tong and Gosele, 1999). The last step in the typical process of substrate wafer fabrication by bonding is thinning of the device wafer by grinding and polishing for a purpose of making its final thickness compatible with the needs of the device manufacturing processes (Fig. 2.23(c)).

The wafer bonding process may produce thin top layers similar to considered earlier epitaxial extension (Fig. 2.21(c)), except that the former is free from constraints regarding lattice matching, and thus, offers greater flexibility in mixing and matching various types of materials for the purpose of accommodating specific needs of any given type of device manufacturing.

Silicon as a substrate for non-silicon materials: Majority of inorganic semiconductors of practical importance discussed in this chapter, whether elemental or compound, can be obtained in the form of the single-crystal wafers that can be used to manufacture semiconductor devices. As discussed earlier, depending on material and single-crystal growth technique such wafers can be circular or square in shape, and may vary in size from 10 mm on the side square wafers in the case of some compound semiconductors, to 450 mm in diameter in the case of circular silicon wafers. In general, larger substrates reduce significantly cost of the devices built upon them and as such are highly desired.

The problem is that because of technology related limitations some semiconductor materials cannot be obtained as the wafers in sizes and at the cost warrantying their usefulness in commercial semiconductor device manufacturing. Considering very high quality of single-crystal silicon wafers, their highly adequate mechanical characteristics, large size, and relatively low cost, silicon wafers are commonly the first choice when it comes to the selection of the alternative hetero-substrates for the technologically challenged semiconductor materials which considered earlier gallium nitride, GaN is a prime example. Formation of thin films of GaN on silicon wafers allows GaN substrates as large as Si wafers and represents economically viable alternative to GaN-on-silicon carbide and GaN-on-sapphire our technologies.

2.8.2 *Non-Semiconductor Substrates*

There are several semiconductor device applications in which substrates in the form of rigid, crystalline, electrically conductive semiconductor wafers discussed in the previous section are not needed, or not desired. In such cases, insulating substrates are employed to provide mechanical support for semiconductor devices built on their surfaces using broadly understood thin-film technology. This section gives a brief overview of insulators used as substrates in semiconductor device engineering.

Sapphire: In the case electronic and/or photonic devices require an insulating substrate featuring adequate optical, mechanical, and chemical characteristics, sapphire is the first choice. Sapphire is a single crystal (hexagonal) form of aluminum oxide Al_2O_3, also known as corundum, which features substrate properties that are highly conducive with the needs of several key electronic and photonic semiconductor devices. Among other insulators, sapphire distinguishes itself with superior resistance to temperature (melting point of 2,300 °C), as well as resistant to aggressive chemistries, and high energy radiation (Kirby, 2007). An added advantage is a high transparency of sapphire (over 80%) to light with wavelengths ranging from about 0.3 μm to about 4 μm. Sapphire is commercially available in the form of the wafers which in terms of diameter and thickness are fully compatible with mainstream semiconductor manufacturing technology. Two different applications of sapphire considered below, demonstrate how it can be used in semiconductor device engineering.

First application of sapphire as a substrate is concerned with Silicon-on-Sapphire (SOS) wafers representing variation of the discussed earlier Silicon-on-Insulator (SOI) technology. In contrast to SOI wafers where the electrically conductive Si wafer is providing mechanical support for buried oxide and active Si layer (Fig. 2.24(a)), in the case of SOS substrates the active Si layer is formed on the bulk sapphire which provides mechanical support and is not electrically conductive (Fig. 2.24(b)). This difference accounts for the significant advantages of the SOS substrates in applications involving devices and circuits built into Si active layer which are designed for operation at very high frequency.

The challenge in the fabrication of SOS wafers is a mismatch between crystal structures of single-crystal Si and sapphire which makes it impossible to grow defect-free, device quality Si active layer on sapphire by means of direct epitaxy. As it turns out, however, density of such epitaxial growth defects can be reduced by amorphization of the defective epi layer by its implantation (see discussion in Chapter 5) with high-energy Si, followed by its reconstruction into a single-crystal phase by means of carefully executed thermal treatments stimulating the process known as solid-state epitaxy (Pohl, 2020).

Second important application of sapphire as a substrate arises from the difficulties imposed by the mentioned earlier difficulties in the processing of forming single-crystal GaN wafers compatible with the needs of large-scale commercial manufacturing of gallium nitride based devices. The heteroepitaxial deposition of GaN on substrates such as sapphire, silicon carbide, or silicon is a solution to this challenge. Among those, sapphire substrate offers lower cost than silicon carbide and transparency to light not available with Si substrates. A transition layer, known as buffer layer, between sapphire substrate and GaN epitaxial layer is needed to accommodate different lattice constants (see earlier discussion of mismatched epitaxial deposition), as well as to minimize the effect of difference in thermal expansion coefficients between these two materials.

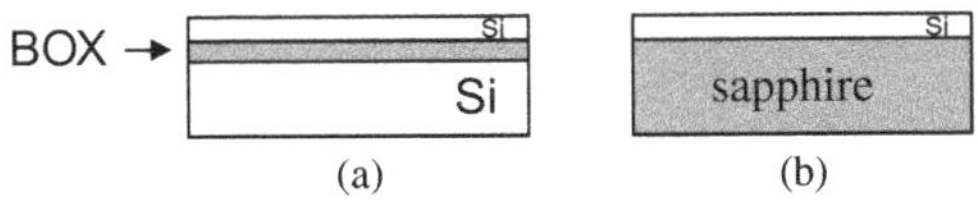

Fig. 2.24. (a) Silicon-on-Insulator (SOI) wafer where buried oxide acts as an insulator and conductive Si acts as a substrate and (b) Silicon-on-Sapphire (SOS) wafer.

Glass: The most common substrate in thin-film technology is glass. It offers insulating properties, transparency to light, and adequate mechanical stability all of which are driving uses of glass as a substrate in semiconductor device technology. The key ingredient of any form of glass is silicon dioxide (SiO_2) known as silica with calcium oxide (CaO) being the second in terms of weight percent component. Additional components including sodium oxide (Na_2O), boron oxide (B_2O_3), aluminum oxide (Al_2O_3), magnesium oxide (MgO), and others are added in various amounts to establish desired properties of glass.

Soda-lime glass, also called soda-lime-silica glass, is the most common type of glass used in our daily lives both in sheet form, e.g. for windowpanes, as well as shaped form, e.g. for glass containers. Due to its very wide use, soda-lime glass, in which second after SiO_2 in terms of weight percent component is nitrous oxide N_2O, accounts for the vast majority of manufactured glass. The soda-lime glass is used as a substrate in semiconductor devices in the case of less demanding in terms of performance and lower cost products. In applications in which resistance to thermal shock is an issue, boron oxide (B_2O_3), instead of lime (N_2O) boron oxide becomes a main component after SiO_2 component resulting in the borosilicate glass featuring superior to soda-lime glass resistance to temperature shocks. Several varieties of this type of glass are available under various trade names such as, for instance, Pyrex™.

In the cases where structural defects inherent to amorphous nature of glass may interfere with characteristics of semiconductor devices formed on its surface, quartz substrates are being used. Quartz, or fused silica is a single-crystal version of SiO_2 which is significantly more expensive but offers superior optical characteristics and resistance to temperature.

In applications in which transparency of the substrate is required, but at the same its surface needs to be electrically conductive, glass slides are covered with the film of material that is at the same time electrically conductive and transparent to visible light. Indium-tin-oxide (ITO) is the most common transparent conductor used for this purpose and ITO covered glass substrates are readily available commercially (see additional discussion of transparent conductive oxides in the following section).

Flexible substrates: Substrates discussed above are mechanically rigid and as such are used only in semiconductor devices and circuits which are not subject to bending, flexing or stretching. When made very thin, essentially any solid, including stainless steel, for instance, will display certain degree of flexibility. In semiconductor device technology, however, bendable and rollable plastic films are the first choice. Among many polymers (*polymer* is essentially another name for plastic) available, choice depends on the degree of flexibility, temperature resistance, stretchability, and transparency. Polyimides, for instance, display good resistance to temperature with Kapton® polyimide tape remaining stable across the temperature range from sub-zero up to 400 °C. A polymer known as PEN (polyethylene naphthalate) is a transparent and conductive material. PEEK (polyether ether ketone) in turn features very good mechanical and chemical resistance characteristics that are retained even at elevated temperatures.

A special in terms of device applications class of flexible substrates are fabrics used to manufacture devices for wearable electronics (see Section 3.9). Properly selected paper is yet another flexible substrate (Kshirsagar *et al.*, 2013) allowing manufacture of thin-film semiconductor devices with electronic paper playing a special role in this regard.

2.9 Thin-Film Insulators

Discussion in this section is concerned with thin-films insulators, featuring thickness varying from 1 nm to some 100 nm depending on application, which are used for various purposes in semiconductor device engineering. No semiconductor device whether electronic or photonic can be fabricated and can operate without an insulator layer being a part of it either in physical contact with semiconductor, with thin-film conductor, or with other thin-film insulator. For these reasons, consideration of the key characteristics of selected thin-film insulators is an integral part of any discussion concerned with semiconductor device technology.

In the discussion of thin-film insulators in this section distinction is being made between multi-purpose thin-film insulators uses of which are not limited to any specific application and dedicated thin-film insulators which are incorporated into device structure to carry out specific function.

2.9.1 General Characteristics of Thin-Film Insulators

As discussed in Section 1.1, insulators are solids which do not conduct electricity. They play pivotal role in semiconductor device engineering as the integral parts of the final device structure where they electrically isolate parts of the device, as well as passivate and protect the surface. In some classes of devices, they comprise the parts upon which device operation is based (see discussion of metal–oxide–semiconductor (MOS) devices in Chapter 3). In some others, insulators play a role in defining optical characteristics of the dielectric-semiconductor structure. Furthermore, thin-film insulators enable semiconductor device manufacturing processes acting, for instance, as masking materials. All in all, no semiconductor device can be fabricated or can operate without thin-film insulators being involved.

In mainstream electronic semiconductor device applications, of interest are insulators which in addition to blocking electric current also do not display permanent polarization and which can get temporarily polarized only in the presence of electric field. Insulators displaying such characteristics are referred to as dielectrics (Baklanov, 2007). It is quite common in everyday semiconductor terminology to consider terms "insulators" and "dielectrics" as synonymous and to use them interchangeably. Furthermore, since many among insulators used in semiconductor technology are oxides (e.g. silicon dioxide, SiO_2, aluminum oxide, Al_2O_3, or hafnium oxide HfO_2), term "oxides" is often used in semiconductor terminology in reference to insulators. (Chang *et al.*, 2007).

An important parameter of dielectrics is dielectric constant k (also known as relative permittivity) which represents insulators ability to store electric charge. Dielectrics featuring high k are capable of holding larger electric charges for longer periods of time than their lower k counterparts. Value of k is a factor upon which various classes of insulators used in semiconductor technology are distinguished.

There are several stringent requirements dielectrics to be used in semiconductor device applications must meet. They are concerned not only with electrical, but also with optical, mechanical, and thermal characteristics of an insulator in contact with semiconductor. In the case of electrical characteristics, high dielectric strength representing resistance of material to the high electric field is of prime interest in some types of devices. In the case of optical characteristics, refractive indices n_1 and n_2 of insulator and semiconductor, defining phenomena of reflection and

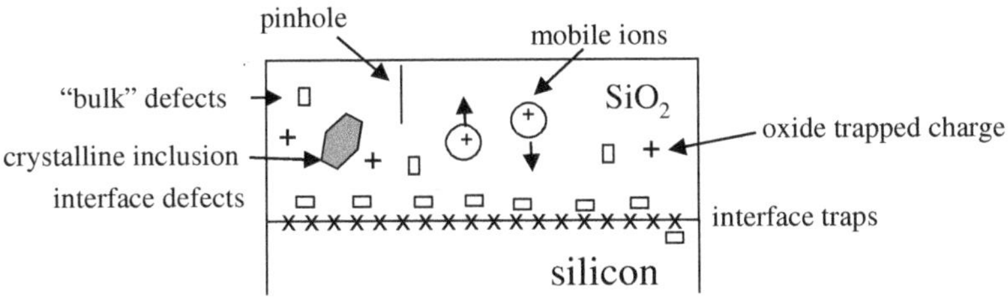

Fig. 2.25. Examples of the defects in the amorphous oxide on single-crystal semiconductor substrate.

refraction of light within the insulator-semiconductor structure, are coming to play. In the case of mechanical and thermal characteristics, adhesion properties of an insulator as well compatibility of thermal expansion coefficients of insulator and semiconductor are important.

Various features of an insulator-semiconductor material system that may have an effect on the stability of its characteristics against electrical and mechanical stress, as well as temperature, time, and high-energy radiation, and in consequence may have an adverse impact on the performance of semiconductor device of which it is a part, are schematically illustrated in Fig. 2.25.

In terms of the process related features, it is important that thin-film insulators can be easily deposited on various substrates involved in semiconductor device fabrication as well as easily removed by etching. Operations employed to process thin-film insulators should be well understood and should not interfere with other materials involved in the manufacture of any given semiconductor device.

2.9.2 *Thin-film Insulators in Semiconductor Device Technology*

A discussion above indicates that only selected insulators feature properties that make them compatible with the needs of semiconductor device technology. The overview below focuses on thin-film insulators which due to their characteristics are used in the range of device related applications irrespective of its function and materials used in its fabrication.

Silicon dioxide (SiO_2): It is the most broadly used insulator in semiconductor engineering not only because it is a native oxide of silicon, but also because it displays advantageous from the device applications point of view characteristics and its deposition and etching technologies are very well worked out. An amorphous silicon dioxide is simply pure glass which, as discussion in earlier section indicates, in the bulk form is a common substrate in semiconductor device technology. In the single-crystal bulk form SiO_2 is known as quartz. In the thin-film semiconductor device applications, SiO_2 is used for various purposes solely in the amorphous form.

As Table 2.4 shows SiO_2 featuring wide energy gap and high dielectric strength is an excellent insulator. It can be readily deposited using available methods and is of particularly high quality when formed on silicon surface by means of thermal oxidation (see Chapter 5). Equally important is an ease with which thin-film SiO_2 can be removed by etching in water solution of hydrofluoric acid, HF. Also, high melting point of SiO_2 defining its resistance to high-temperature treatments is yet another property defining its usefulness in the manufacture of the range of semiconductor devices.

A shortcoming of SiO_2 is its relatively low density (Table 2.4) which makes it permeable by moisture and hydrogen as well as alien mobile ions, alkali ions such as sodium Na^+ in particular (Fig. 2.25). Once in SiO_2, contaminating ions can move around under the influence of the electric field inflicting instabilities of the device incorporating thin SiO_2 film. Second shortcoming of silicon dioxide is its sensitivity to high-energy radiation, or in other words insufficient radiation hardness. This last deficiency of SiO_2 is causing problems when devices and circuits

Table 2.4. Characteristics of selected thin-film insulators. *(Values shown are for comparison purposes only and may vary depending on the thin-film deposition method)*

Property /Material	Silicon dioxide SiO_2	Silicon nitride Si_3N_4	Aluminum oxide Al_2O_3
Energy gap width E_g (eV)	8.0	5	8.4
Resistivity (10^{15} Ω-cm)	10	0.1	10
Dielectric constant, k	3.9	5	8
Density (g/cm^3)	2.25	3.4	3.8
Refractive index	1.46	2.02	1.7
Dielectric strength (MV/cm)	10	10	3
Thermal conductivity (W/cmK)	0.014	0.3	0.3
Melting point (°C)	1,700	1,900	2,000

incorporating SiO_2 are used in space and military applications. Deficiencies of SiO_2 can be to some degree alleviated by adding nitrogen to SiO_2 and forming silicon oxynitride, SiO_xN_y. The procedure results in the increased density of an oxide and improvement of its overall integrity including radiation hardness.

Silicon nitride (Si_3N_4): Another insulator often used in semiconductor devices is Si_3N_4. While a compound of silicon, silicon nitride cannot be formed on Si surface by reaction of silicon with nitrogen, and thus, can only be deposited through the reaction of silicon and nitrogen taking place in the gas phase. A good insulator, Si_3N_4 features higher density than SiO_2 (Table 2.4) which makes it less vulnerable to the penetrations by alien elements, including oxygen, even at high temperature. The downside of this characteristics is that Si_3N_4 is significantly more difficult to etch than SiO_2. It also features inherent to its structure defects which may have an adverse effect on the devices incorporating Si_3N_4. Furthermore, silicon nitride forms low-quality interface with silicon, and hence, in spite of being a silicon compound is not used in conjunction with Si in the case when low density of the electric charge of the system are required.

Aluminum oxide (Al_2O_3): Similar to silicon nitride role in semiconductor device technology is being played by amorphous thin-film Al_2O_3 which in the single-crystal form is known as sapphire. Displaying comparable to silicon dioxide and nitride insulating properties, amorphous aluminum oxide, also referred to as alumina, features higher density (Table 2.4), which makes it a strong barrier against moisture and contamination. It also features very good thermal characteristics in terms of the melting point, thermal conductivity, and thermal expansion coefficient which, in conjunction with very high chemical resistance, make aluminum oxide a particularly rugged insulator.

Applications dependent on dielectric constant: Besides thin-film insulators capable of performing various functions discussed above, there is a class in which are selected for specific application based either on their inherent characteristics such as high or low value of dielectric constant *k*, or their ability to change some of their physical properties upon modifications of their chemical composition.

Specific needs of semiconductor device engineering often call for the choices in terms of insulator material properties which are conducive with the needs of the device in which they are incorporated. Common selection criterion is based on the value of dielectric constant *k* of the material. As mentioned earlier, dielectric constant, *k*, is a parameter defining ability of material to store charge. Consequently, it also defines capacitance *C* of any capacitor comprising of the layer of dielectric sandwiched between two conducting plates. All other parameters equal, *k* determines capacitance of such structure, or in other words, it defines the extent of capacitive coupling between two conducting plates. With dielectrics featuring high value of *k* such coupling is strong while with those featuring low *k* value, coupling between two conducting materials separated by such dielectric is obviously weak.

In semiconductor terminology rough distinction between dielectric materials is based on the value of *k* of silicon dioxide, SiO_2, which as Table 2.4 shows is 3.9. Dielectrics featuring $k < 3.9$ are commonly referred to as low-*k* dielectrics. For an insulator to perform in practical devices a function of a high-*k* dielectric and to be referred to as such, its dielectric constant needs to be in $k > 15$ range. In cutting edge semiconductor electronics both high- and low-*k* dielectrics are needed to implement fully functional devices and circuits. Insulating materials displaying adequate *k* value and meeting requirements of semiconductor device technology are as follows.

High-*k* dielectrics: In general, dielectrics featuring high dielectric constant *k* (high-*k* dielectrics), are needed to increase capacitive coupling. As discussed in more details in Chapter 3, high-*k* dielectrics are needed to assure sufficiently high capacitance of the Metal-Oxide (dielectric)-Semiconductor structure in nano-scale MOS/CMOS transistors. While in some other devices dielectrics featuring value of *k* in the range of 100 and higher may be desired, in MOS gate configuration such high values of *k* may bring about deleterious fringing effects hampering transistor's performance. As a result, in applications concerned with thin-film dielectrics used in cutting-edge MOS transistors somewhat more moderate values of *k*, in the range from about 20 to about 50, are of interest. This includes amorphous hafnium dioxide, HfO_2 (hafnia) featuring $k \sim 25$ in particular, and zirconium oxide, ZrO_2 (zirconia) featuring similar *k* to a lesser degree because of thermal instabilities of its structure. In addition, hafnium silicate and zirconium silicate, $HfSiO_4$ and $ZrSiO_4$ respectively, which feature adequate properties, including thermal stability, but lower than needed in typical application value of *k* (below 20) are suitable for some high-*k* dielectric applications.

Other than the use in transistors, high-*k* dielectrics are also needed in memory devices in which high capacitance of the storage capacitors at the limited capacitor area is important. In certain types of memories, dielectric materials featuring moderate *k* such as aluminum oxide (Table 2.4) and tantalum oxide are sufficient, but in most others, dielectrics featuring significantly higher *k* are desired.

Depending on the type of memory cell, dielectrics featuring *k* in the 80–100 range, such as titanium dioxide (TiO_2) or complex oxides built on the titanium oxide framework and featuring ferroelectric properties are used. Ferroelectrics feature spontaneous polarization in contrast to dielectrics which require electric field to get polarized. Many of such ferroelectric complex oxides belong to the class of oxides which unlike amorphous dielectrics discussed earlier, feature perovskite crystal structure. The common feature of this class of solids is a very high dielectric constant *k* ranging in hundreds and exceeding 1,000 in some cases.

As name indicates, complex oxides, alternatively known as functional oxides, are multi-element materials which besides oxygen include from two to five other elements and which, as mentioned earlier, are predominantly built on the titanium oxide framework. Examples of compounds representing the class of ferroelectrics featuring perovskite structure (Baklanov *et al.*, 2007) include lead zirconate titanate, known in short as PZT, as well as lead-free barium strontium titanate, BST. Within the same class of complex oxides are materials displaying piezoelectric characteristics allowing conversion of the mechanical stress into electrical signal. Piezoelectric materials are often integrated with semiconductor micro–electro–mechanical systems (MEMS) discussed in Section 3.8 of this volume.

While the potential of complex oxides in semiconductor device engineering is being recognized, more detailed discussion of the physical phenomena determining characteristics of a broad class of ferroic materials, including complex oxides, is beyond the scope of this brief overview.

Low-*k* dielectrics: On the opposite to high-*k* end of the spectrum, there are applications in which as weak as possible capacitive coupling, limiting as much as possible a cross-talk between electrically isolated neighboring conducting lines, must be assured. In such cases insulating materials featuring dielectric constant *k* as low as possible, which translates to the *k* values lower than 3.9, and as close as possible to 1(dielectric constant of air) need to be applied. Situation of this nature occurs in multi-layer metallization schemes implemented in advanced integrated circuits (see Chapter 3) in which low-*k* inter-layer dielectrics (ILD) are used to electrically insulate metal lines. In the most advanced integrated circuits operating at high frequencies, *k* of an ILD below 2.0 is desired.

Reduced *k* of an insulator can be accomplished in three different ways. First, reduction of the *k* value can be achieved through modifications of the chemical composition of material. For instance, addition of fluorine (F) into silicon dioxide (SiO_2) results in the reduction of *k* from 3.9 to about 3.0 depending of the process employed. Second option, which delivers *k* in the range from about 3.0 to about 2.0 depending on the dielectric material, is to employ selected organic solids to act as a low-*k* dielectrics. Among them selected polyimides, polymers, carbon-based compounds, as well as teflon may serve this purpose. The third way to reduce dielectric constant of a dielectric is in introduce porosity to its structure. Considering *k* of the air being 1, inclusion of the air gaps into the material such as SiO_2, for instance, may lower its *k* to the value approaching 1. At such high porosity, mechanical cohesiveness of the thin-film dielectric is an issue which needs to be addressed.

2.10 Thin-Film Conductors

In addition to insulators, electricity conducting solids, or conductors, are indispensable in the technology of semiconductor devices. From the point of view of basic physical properties the difference between conductors, insulators, and semiconductors were discussed in Chapter 1. The role of conductors in semiconductor device technology is to provide low resistance contacts allowing electric current to flow in and out of the device as well as to form very high conductivity lines interconnecting devices within an integrated circuit. It needs to be stressed that in the case of contacts, selection of the conductor used depends on the semiconductor material with which contact is made (see discussion of ohmic contacts and Schottky contacts in the following chapter). Furthermore, in semiconductor devices conductors are applied only as thin films which emphasizes their material properties specific to thin-film configuration.

This section briefly reviews most important materials used as conductors in semiconductor device applications including metals, metal alloys, and non-metallic conductors.

2.10.1 *Metals and Metal Alloys*

Under normal conditions metals feature the highest electrical conductivity among solids and as such are broadly employed as conductors in semiconductor devices and circuits. To be of use in this very demanding application, however, in addition to being excellent conductors of electricity, metals must meet other stringent requirements. Among them the following are particularly critical: (i) no changes of the key physical properties in the presence of the high density current including uncontrolled migration of metal's atoms referred to as electromigration, (ii) easy deposition and easy removal by etching, (iii) no uncontrolled chemical/electrochemical interactions with materials in physical contact, (iv) good adhesion to various solids in device structure, (v) resistance to elevated temperature, and (vi) structural homogeneity showing no grains or other imperfections.

There is no single metal element in the Periodic Table which fully meets all the requirements listed. However, as the requirements listed are not equally critical in all applications, there are metals available which perform sufficiently well in some specific applications while being not suitable in some others (Wetzig and Schneider, 2003). Table 2.5 lists few among those which are used in semiconductor device engineering and shows their deficiencies which are important from the device technology perspective. Among metals finding applications in semiconductor device technology copper (Cu) features the lowest electrical resistivity/highest electrical conductivity, but because of the shortcomings listed in Table 2.5 its use is limited to the formation of the interconnect lines in advanced integrated circuits (see Chapter 5).

Historically, aluminum (Al) is the most commonly used thin-film metal in semiconductor device technology. It features high electrical conductivity, can be easily deposited and etched and under normal conditions is chemically neutral with respect to semiconductors of interest. Due to all this, it still remains important in the range of device applications, but its usefulness in some others is limited due to either electromigration or limited temperature resistance, or both.

Table 2.5. Characteristics of selected metals used in semiconductor technology. *(at 300 K; values are shown for comparison purposes and may vary depending on the thin-film deposition method)*

Material /Property	Resistivity ($\mu\Omega$-cm)/	Melting point (°C)	Deficiencies
Copper, Cu	1.7	1,084	high chemical reactivity, contaminant of silicon, difficult to deposit/etch
Gold, Au	2.3	1,063	difficult to etch, contaminant of silicon
Aluminum, Al	2.7	660	electromigration, low temp. resistance
Tungsten, W	5.6	3,422	difficult to deposit/etch

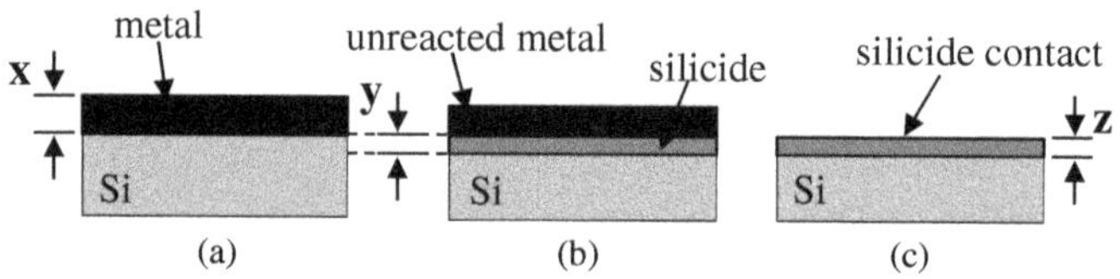

Fig. 2.26. Process of silicide formation, (a) metal deposition, (b) sintering, and (c) removal of unreacted metal.

In addition to aluminum, and despite the cost and process challenges (difficult to etch), gold (Au) is commonly used as a contact material in various compound semiconductors. Not in silicon, however, which can be readily penetrated by gold causing formation of defects severely affecting charge carriers transport in silicon.

Finally, tungsten (W), featuring the highest of all metals melting point of 3,422 °C represents refractory metals known not only for extraordinary temperature resistance, but also for their hardness and corrosion resistance. It plays an important role of the via material in the interconnect technology (see Chapter 4).

To improve their characteristics and versatility, some metals are processed to form alloys or are chemically modified to make them compatible with the needs of particular process or device. Most common are alloys of silicon with selected metals to form metal silicides. They are commonly used in Si devices to form ohmic contacts to silicon. As Fig. 2.26(a) shows, first step in the process is deposition on silicon surface of thin film of metal featuring thickness x. Subsequently, temperature is increased to the point where silicon and metal phase are mixing forming an alloy. The process is commonly referred to as sintering. Temperature at which it occurs is known as sintering temperature and is an important characteristic of any given silicon-metal alloy system. Another process defining characteristic is the depth of metal penetration into silicon during sintering process indicated as y in Fig. 2.26(b). In the last step of the silicidation process the unreacted metal remaining on top of the silicide is removed by etching leaving behind a layer of silicide featuring thickness z (Fig. 2.26(c)).

In general, silicides as ohmic contacts to silicon are of interest because of (i) ease of patterning on Si surface (see self-aligned process known as *salicide* process discussed in Chapter 5), (ii) superior resistance to temperature as compare to low-resistivity metals such as aluminum or gold, and (iii) comparable to or lower resistivity than temperature hardy, but difficult to process refractory (high melting point) metals including tungsten.

Several metals form alloys with silicon. The first order criteria for the selection of metals to form a silicide are based on the resistivity of silicide, sintering temperature (lower the better), and depth of alloy penetration into Si, y in Fig. 2.26(b) (in general, shallower the better). Frequently used silicides include nickel silicide (NiSi) featuring resistivity of about 16 $\mu\Omega$-cm, lowest among other silicides, sintering temperature in the range of 500 °C, and shallow penetration of silicon during silicidation process. Among others, titanium silicide ($TiSi_2$) features resistivity in the range of 15 $\mu\Omega$-cm, but higher than NiSi sintering temperature of about 800 °C. It also penetrates silicon deeper than NiSi. Comparable to $TiSi_2$ characteristics are displayed by cobalt silicide ($CoSi_2$).

As mentioned earlier, it is common to add alien elements to any given metal to improve its performance in specific device applications. For instance, nitrogen is often added to titanium to form titanium nitride TiN (known as "tinitride") which is one of the most common metal alloys in semiconductor engineering. In addition to titanium, other refractory metals, tantalum, Ta, and tungsten, W, form alloys with nitrogen (TaN, WN) useful in contact applications. Other than that, small amounts of silicon can be added to aluminum and some other metals to modify properties toward better compatibility of a given metal with some semiconductor devices.

2.10.2 *Non-Metallic Conductors*

While decidedly the most electrically conductive, metals are not the only electrical conductors used in the construction of semiconductor devices.

In certain situations, very high electrical conductivity (low resistivity) of metals is being compromised on behalf of the benefits with respect to the performance of the device resulting from the use of selected non-metallic conductors. Two examples of such non-metallic conductors include poly-crystalline silicon and transparent conducting materials.

Polysilicon. As discussion in the following chapter will reveal, electrically conductive polycrystalline silicon, or poly-Si in short, is a desired thin-film contact material in certain types of single-crystal silicon based transistors due to the similarity of work function between these two crystallographic forms of the chemically same material. The difference is in the resistivity of polysilicon conductors, which is typically in the range of 800 $\mu\Omega$-cm, and which is few orders of magnitude lower than the resistivity of conventional single-crystal Si used to fabricate functional devices.

Very low electrical resistivity is not an inherent characteristic of poly-Si. If needed, its low resistivity is accomplished by the process of doping considered in this book. Dopant atoms are introduced into poly-Si in high concentrations either during the film deposition, or after completion of the deposition process.

Transparent conducting materials: Operation of some important semiconductor devices requires undisturbed interactions with light getting either in or out of the device. Even if kept ultra-thin, i.e. in the thickness regime below 10 nm, metals may not be sufficiently transparent to light in the visible part of the spectrum to assure adequate operation of such devices. This limitation mostly precludes application of metals as electrical contact materials in some classes of photonic devices which are based on either absorption or emission of light. Instead, thin-film materials which are good conductors of electricity and at the same time are transparent to light by virtue of wide energy-gap E_g are commonly employed as contact materials.

A special role of electrical contacts in transparent electronics and photonics is played by transparent conductive oxides (TCOs) among which mentioned earlier indium tin oxide (ITO) with molecular formula indium tin oxide (ITO) is the most commonly used TCO in commercial device applications. Its properties can be manipulated by adding dopants such as fluorine resulting in fluorine doped tin oxide (FTO). Another transparent conductive oxide is zinc oxide, ZnO. In spite of its oxide denotation that is typically associated with insulating materials, ZnO is a group II–VI compound semiconductor discussed earlier in this chapter. Similarly to ITO, chemical makeup of ZnO can be altered by doping with, for instance, aluminum to modify its electrical and optical characteristics.

In addition to TCOs, also selected conductive polymers with composition manipulated towards achieving 90% or better transparency to visible light are used as transparent conductive materials. Materials of this type are based on polyacetylene, polyaniline, and other established classes of polymers.

Besides mechanically coherent thin films of transparent conductive materials, transparency and high electrical conductivity can be accomplished by using nanoscale material system such as 1D carbon nanotubes, or 2D graphene discussed in Section 2.3.

Review Questions

2.1. Explain the difference between single-crystal material, polycrystalline material and amorphous material.
2.2. Consider cubic crystal cell. Using Miller indices define (100), (110), and (111) crystallographic planes.
2.3. Each semiconductor device, in the text above referred to as semiconductor device material system, includes components constructed using various types of materials. Identify them and define purposes they are used for.
2.4. What elements included in Periodic Table of Elements feature semiconductor properties? Among them which one is the most often used to manufacture semiconductor devices and why?
2.5. Name two physical characteristics of the conventional inorganic semiconductor materials deciding in what type of device they are used.
2.6. Identify binary compounds formed using elements from group IV of the periodic table which are commonly used in power applications.
2.7. Give example of the ternary semiconductor material.
2.8. Identify three 2D materials featuring semiconductor properties.
2.9. What would you say is the property of organic semiconductor materials determining their various applications?

2.10. Describe CZ single-crystal growth method.
2.11. Assume you need to deposit thin film of single-crystal semiconductor material on single-crystal substrate. Name the method you suggest using and describe its workings.
2.12. What semiconductor and non-semiconductor substrates are used to manufacture functional semiconductor devices?
2.13. What is a SOI substrate and describe possible ways of its fabrication.
2.14. List dielectric materials (insulators) most often used in semiconductor device technology.
2.15. What metals in the form of thin films are used in semiconductor device technology and for what purposes?

Chapter 3

Semiconductor Devices

3.1 Introduction

As a class of materials, semiconductors play an undeniably pivotal role in the explosive growth of our technical civilization over the last several decades. The main driving force behind this growth was the unprecedented progress in integrated circuits (IC) technology.

During recent years, the progress in integrated circuits technology is accompanied more visibly than before by the accelerated growth of distinct, readily identifiable semiconductor technical domains which are not directly related to IC technology but which since early years of semiconductor engineering were considered in literature and applied in industrial practice (Brophy, 1964; Grove, 1967; Cooke, 1990; Enderlein and Horing, 1997; Streetman and Banerjee, 1998; Neamen, 2011; Burghartz, 2013).

The goal of this chapter is to identify major classes of semiconductor devices, discuss principles of their operation, and show how they contribute to the growth of semiconductor electronics and photonics by considering their main uses.

Discussion in this chapter starts with an overview of various classes of semiconductor device followed by review of the principles upon which semiconductor devices are constructed. Subsequently, two-terminal semiconductor devices, diodes, and three-terminal devices, transistors are considered.

Various types of semiconductor diodes designed for diverse electronic and photonic applications are identified. Then, various classes of transistors are discussed with the role of MOS field-effect transistors (MOSFET) forming building blocks of the Complementary MOS (CMOS) cells emphasized. Also, thin-film transistor (TFT) version of the MOSFET is considered. A separate section is devoted to the overview of integrated circuits (ICs) which are at the core of the electronic systems and consumer products used in our daily lives.

In the remaining parts of this chapter semiconductor imaging devices (image displaying and image sensing) are considered as well as micro–electro–mechanical systems (MEMS) exploiting mechanical characteristics of silicon are discussed. The last section is devoted to a brief overview of the wearable and implantable semiconductor device systems.

Interweaved in the discussion in this chapter are remarks specifically addressing the uses of semiconductor devices in everyday consumer and industrial products. The purpose here is to demonstrate how deeply our lives in 21st century are dependent upon semiconductor science and engineering.

3.2 Overview of Semiconductor Devices

Semiconductors and other materials discussed in Chapter 2 do not serve useful purpose unless they are engineered into functional devices. The term "semiconductor device" is used here in reference to a piece, or a thin-film of semiconductor material, combined as needed with thin layers of insulators and conductors, all configured in such way that the resulting material system can perform in the controlled fashion predetermined electronic, photonic, or electro-mechanical functions (Sze, 1985, 1990, 1998; Pierret, 1996; Sze *et al.*, 2021).

Electronic functions are performed by electronic devices which operation is based on the interactions of electric charge carriers and in which electrons are acting as information/energy carriers (Kasap, 2006). Term photonic devices is concerned with devices involving interactions of photons ("packets" of electromagnetic energy $h\nu$ carried by light) and in which photons are acting as information/energy carriers.

Electro-mechanical devices in turn are taking advantage of the mechanical characteristics, such as elasticity and fracture toughness of some semiconductor materials, silicon in particular. In contrast to electronic and photonic functions, which involve interactions solely within semiconductor material systems, electro-mechanical functions require

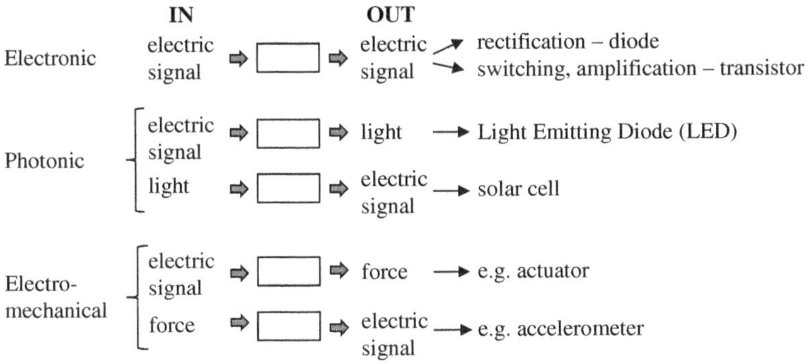

Fig. 3.1. Various classes of semiconductor devices distinguished based on the input and output signal.

involvement of solids capable of converting mechanical action into electrical signal (and vice versa) such as piezoelectrics.

Figure 3.1 illustrates schematically principles upon which three types of semiconductor devices defined above operate. In the case of electronic devices both input and output signals are electrical and the purpose of such devices is to process electrical signal. In the case of photonic devices two types of interactions, and thus two classes of devices, are distinguished. In the first one, device converts input electrical signal into light while in the second, opposite action is implemented. Completing the list in Fig. 3.1 are electro-mechanical devices built into semiconductor material. They are designed to react mechanically to stress inducing force which is then converted by integrated piezoelectric material into electrical signal. Action in the opposite direction can also be implemented.

The following discussion covers fundamentals defining basic types of semiconductor devices, describes in general terms principles of their operation, and considers selected aspects of device configuration. The discussion is qualitative in nature and, in agreement with the scope of this book, is not involving elements of semiconductor physics underlying device operation beyond basic concepts introduced in Chapter 1. Furthermore, it does not attempt to address ever-expanding diversity of semiconductor device designs and functions. Instead, the field of semiconductor devices considered here is limited to the overview of those devices both discrete and integrated, which represent current and emerging trends in semiconductor electronics and photonics.

3.3 Constructing Semiconductor Device

There are two fundamental elements that need to be included in the process of converting semiconductor material into functional device. First, to accomplish device's ability to control flow of current passing through its body a potential barrier must be built into the device structure. Also, it needs to be assured that the electric current can flow in and out of semiconductor material system comprising a device in the undisturbed fashion. To accomplish this task, ohmic contacts need to be formed at the device input and output. Concepts of potential barriers and ohmic contacts constituting building blocks of semiconductor devices are considered in this section.

3.3.1 *Ohmic Contacts*

In considered here ohmic contacts, term *ohmic* comes from the name of the physicist Georg Ohm, (1789–1854) and refers to the electrical contact between metal and semiconductor featuring very low resistance (Fig. 3.2(a)). A key role of an ohmic contact is to allow undisturbed in any way flow of the current in and out of the device regardless of the direction of the applied voltage. In this way connection between device and outside circuitry is not interfering with device operation.

Under such conditions potential V along the length of the sample (Fig. 3.2(b)) is constant. Current flowing across the device changes with bias voltage as shown in Fig. 3.2(c), but under constant conditions in terms of temperature and illumination, resistance of such device does not depend on the applied voltage regardless of direction in which it is applied. This feature manifests itself in the constant slope of the I–V (current–voltage) curve shown in Fig. 3.2(c).

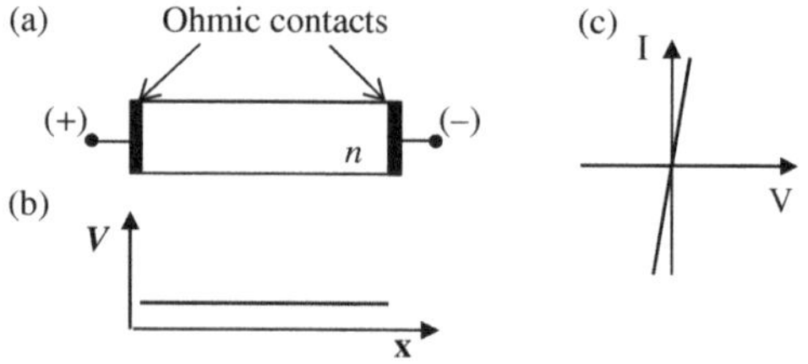

Fig. 3.2. (a) A piece of semiconductor material equipped with ohmic contacts, (b) undisturbed, uniform distribution of the potential along semiconductor with ohmic contacts, (c) fully symmetric output current–voltage characteristics.

To form an ohmic contact to any given semiconductor, selection of an adequate metal is essential. The goal is to select a metal to form a contact with semiconductor such that the work functions (see Fig. 1.5) of these two solids are the same (or at least very similar). Only then, Fermi levels in the two materials in physical contact are aligned, and flow of carriers from metal to semiconductor and from semiconductor to metal is undisturbed.

Ohmic contacts constitute an integral part of essentially any type of semiconductor device identified in Fig. 3.1 with a role to assure undisrupted flow of the electric current in or out of the device. It needs to be pointed out, however, that even with the matched work functions between metal and semiconductor, electric contact may still not perform up to the needs. This is because any material or process malfunction encountered in the formation of ohmic contact will adversely alter its I–V characteristics (Fig. 3.2(c)) in terms of their linearity, symmetry, and reduced slope, all of which may render device dysfunctional.

All in all, processing of the metal-semiconductor contacts which display ohmic characteristics is not a trivial issue and requires attention to every detail of the fabrication process. The result should be such that the ohmic contacts are "transparent" to the flowing current which means that any changes in the value of the output current should only be due to the changes in the distribution of potential in its body and without any modification of the current flow caused by the contacts.

3.3.2 *Potential Barrier*

In somewhat general terms, operation of semiconductor device is based on the ability to control the current flowing across it by either varying concentration of free charge carriers available for conduction, or by creating a potential barrier affecting flow of charge carriers. In the device shown in Fig. 3.2 current increases linearly with applied voltage, but its resistance is not changing because concentration and distribution of charge carriers is not changing. Equipped with ohmic contacts, piece of semiconductor in Fig. 3.2(a) acts as a simple resistor displaying symmetric, independent of the direction of applied voltage I–V characteristics (Fig. 3.2(c)). Shining light with energy $E > E_g$ on the device in Fig. 3.2(a) at the set bias voltage, increases concentration of free charge carriers, and thus decreases its resistance, but the symmetric character of the I–V characteristics remains unchanged. Under such conditions the device acts as photoresistor. Similar effect is observed when temperature of the device is increased enough to generate additional free charge carriers causing increase of the device current and having it act as a device known as thermistor.

While useful as light or temperature sensors, neither photoresistor nor thermistor are able to perform as devices in which direction of applied bias voltage alters current flowing across its terminals. To accomplish voltage-controlled semiconductors devices with non-linear, non-symmetric current–voltage (I–V) characteristics, a potential barrier must be formed in the piece of semiconductor material equipped with ohmic contacts. The applied voltage dependent height of the potential barrier can then be used to control device current.

The potential barrier can be formed within semiconductor device in various ways and its origin and the way in which it is used determines not only electrical characteristics of active semiconductors devices but also defines classes to which each device belongs.

In general, a potential barrier is formed when semiconductor is brought to physical contact with other material featuring different work functions. Options in this regard include contact between semiconductor and semiconductor featuring different work functions, or contact between semiconductor and metal, or other conductor, featuring different work function. The third option involves contact between semiconductor and insulator in which case, however, the very presence of non-conductive insulator rather than potential barrier resulting from the work function difference between materials in contact determines current–voltage characteristics.

The most obvious way to create a potential barrier in semiconductors is to bring into contact two semiconductors with different work functions. These may be two pieces of the same material such as silicon, providing however, each of them is doped at the different level and/or feature different conductivity type (*p*-type semiconductor and *n*-type semiconductor), and thus, feature different work function. Schematic diagrams in Fig. 3.3 show how the potential barrier is formed in the piece of semiconductor in the absence of the voltage bias when *p*-type and *n*-type semiconductors are brought into contact to form a structure known as a *p–n* junction.

After *n*- and *p*-type materials are brought into contact (Fig. 3.3(a)), a trend to even out large concentration gradients (high concentration of electrons in *n*-type part of the junction and holes in the *p*-type part of the

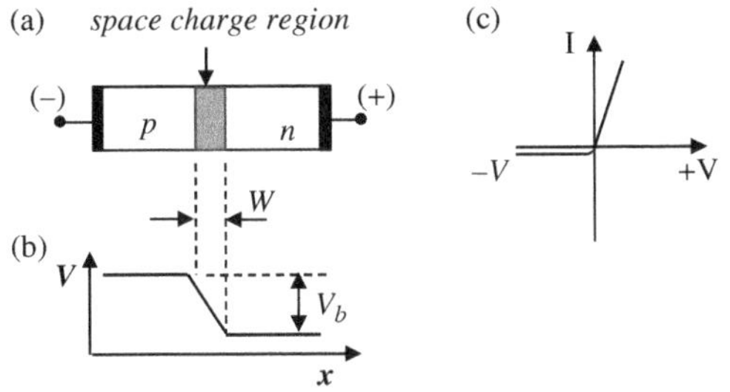

Fig. 3.3. (a) The *p–n* junction device, (b) potential barrier created at the junction, (c) resulting rectifying current–voltage (I–V) characteristic.

junction) will enforce the flow of electrons from *n*- to *p*-type regions and holes in the opposite direction. The left behind immobile donor and acceptor ions form the space-charge regions near the junction with resulting electric field build up eventually preventing further flow of carriers. To maintain thermal equilibrium ($np = n_i^2$) across the junction, energy levels at its both sides will be realigned such that *n*- and *p*-type parts will feature different potential resulting in potential difference across the junction the creation of the potential barrier V_b (Fig. 3.3(b)). The width of the formed junction, W, corresponds to the width of the space-charge region in which an electric field established in the process of junction formation sweeps it of any free charge carriers.

The point of these considerations is that once potential barrier is formed, its height V_b and width W of the space-charge region can be changed by the voltage applied to the junction. Reverse bias, $V < 0$ applied to the *p*-type region, increases potential barrier height preventing majority carriers from flowing across the junction and allowing only small minority carrier current to flow (Fig. 3.3(c)). Control of the reverse current is possible only up to certain value of the reverse voltage beyond which breakdown of the *p–n* junction occurs resulting in the unrestricted flow of current across the junction. Forward bias, $V > 0$ applied to *p*-type region, reduces barrier heights and allows large majority carrier current to flow across the junction (Fig. 3.3(c)). The resulting non-symmetrical, rectifying current–voltage characteristics of the *p–n* junction shown in Fig. 3.3(c) indicate its diode-like behavior with the possibility of wide range applications in electronic circuits. In these applications, *p–n* junction based devices are referred to as bipolar devices because both majority and minority carriers contribute to the flow of current across the junction.

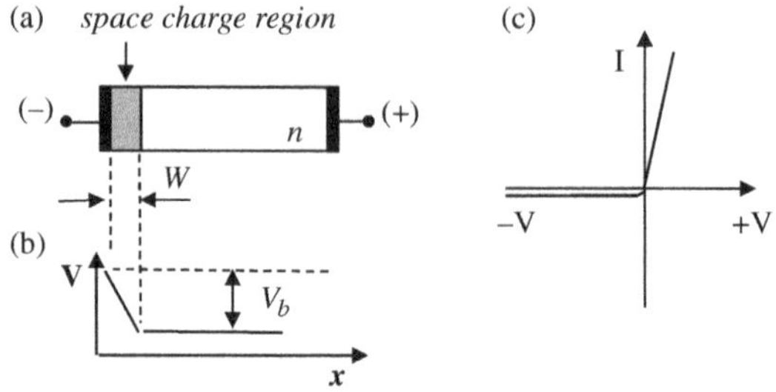

Fig. 3.4. (a) Schottky metal-semiconductor contact, (b) potential barrier created at the metal-semiconductor contact, (c) resulting rectifying current–voltage (I–V) characteristic.

An alternative to *p–n* junction way of potential barrier formation is by bringing into contact semiconductor and metal featuring different work functions and to form the metal-semiconductor contact (Fig. 3.4(a)) also known as Schottky contact (after physicist Walter Schottky, 1886–1976). As shown in Fig. 3.4(b), also in this case difference in the work functions of materials in contact alters potential distribution across the device, and results in establishing of the potential barrier V_b and the space-charge region featuring width W, at the Schottky contact (compare potential distribution in the device in Fig. 3.2 featuring two ohmic contacts and device in Fig. 3.4 featuring Schottky contact and ohmic contact). Similarly to *p–n* junction, a reverse bias voltage applied to the Schottky junction ($V < 0$ on metal in the case of *n*-type semiconductor) increases height of the potential barrier and prevents the flow of majority carriers from semiconductor to metal (Fig. 3.4(c)). A forward bias lowers the potential barrier at the contact and allows majority carriers current to flow from semiconductor to metal. Similarly to the *p–n* junction, the result is a non-symmetrical, rectifying current–voltage characteristic shown in Fig. 3.4(c).

Because only majority carriers are involved in the operation of device shown in Fig. 3.4, Schottky contact based devices are referred to as unipolar devices. Additional difference between *p–n* and metal-semiconductor junctions is that in the latter case barrier heights is controlled not only by the work function difference between metal and semiconductors, but also by the density of the surface states associated with a disturbed crystal structure of the semiconductor surface in contact with metal. Another difference, discussion of which is beyond the scope of these considerations, is a different mechanism of charge carrier's transport over the potential barrier in the case of *p–n* junction and Schottky contact.

The third way of potential barrier formation involves bringing semiconductor into contact with an insulator which results in the alteration of potential distribution in semiconductor in the region immediately adjacent to its interface with insulator. To convert such structure into a current controlling device, metal contact needs to be formed on the surface of an insulator converting into what is known as Metal-Insulator-Semiconductor, or MIS configuration (more commonly used is synonymous term Metal-Oxide-Semiconductor, or MOS) with ohmic contact formed on the back surface of semiconductor (Fig. 3.5(a)).

The difference between *p–n* junction and Schottky contact structures at one end and MOS device on the other lies in the insulator thickness x_{ox} in Fig. 3.5(a) controlling current flow across MIS device rather than the height of the potential barrier as in the former two structures. In the case an insulator x_{ox} is ultra-thin (thinner than about 3 nm), current may flow between metal and semiconductor across the insulator by means of the tunneling, and such MOS structures are called accordingly MIS (MOS) tunnel devices. Tunneling is a charge carriers transport mechanism allowing electron to cross potential barrier associated with an insulators sandwiched between metal and semiconductor without changing its energy (Fig. 3.5(b)). Non-symmetrical I–V characteristic similar to this of the *p–n* junction (Fig. 3.3(c)) and Schottky contact (Fig. 3.4(c)) results, but the one displayed by the MIS (MOS) tunnel devices do not feature satisfactory rectifying properties to be of use in practical applications.

With an insulator (oxide) sufficiently thick to prevent tunneling current between two conductors, the structure shown in Fig. 3.5(a) assumes characteristics of the capacitor and from the applications point of view

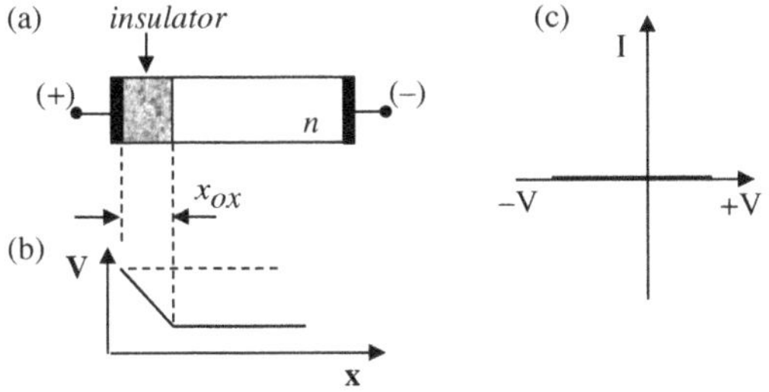

Fig. 3.5. (a) Insulator – semiconductor material system, (b) change in the potential reflects voltage drop on the layer of insulator, and (c) no current flows across the insulator regardless of the bias.

represents entirely different than tunnel structure class of semiconductor devices. It is not possible to precisely define oxide thickness x_{ox} in MOS structure at which changes from diode-like to capacitor characteristics occur, because it depends on the number of variables, including selection of materials comprising MOS structure and the bias voltage. For the sake of this discussion, however, it can be assumed that the oxide thickness in the range 3–4 nm represents thickness regime in which distinct differences between diode-like behavior of the MOS structure with ultra-thin oxide, and capacitor-like characteristics of the MOS structure with thicker oxide are coming to play. From this perspective MOS tunnel devices can be seen as intermediate structures between Schottky contacts (oxide thickness $x_{ox} = 0$), and MOS capacitors (oxide thickness $x_{ox} > 3$ nm) which are further considered later in this section.

As the discussion below will reveal, the ways of potential barrier formation discussed above, namely *p–n* junction, Schottky contact, and MOS structure lay a foundation upon which active semiconductor devices are formed.

3.4 Diodes

The term *terminal* refers to the point in the electronic circuit in which connection of the device to the external circuit is established. Devices with two terminals represent the simplest, very important in many regards class of semiconductor devices.

The two-terminal devices featuring rectifying current–voltage (I–V) characteristics discussed in the previous section are referred to as diodes. Accordingly, *p–n* junction diodes and Schottky diodes are distinguished. In addition, the Metal-Oxide-Semiconductor (MOS) capacitors complete a family of two-terminal semiconductor structures which constitute a self-contained class of semiconductor devices, but which also serve as a basis for the development of the three-terminal devices known as transistors.

3.4.1 *Electronic Diodes*

The discussion of semiconductor diodes in this section is focused primarily on the *p–n* junction diodes with reference to Schottky diodes in relation to specific applications only. This approach is justified as the former are the most important in practical applications semiconductor diodes.

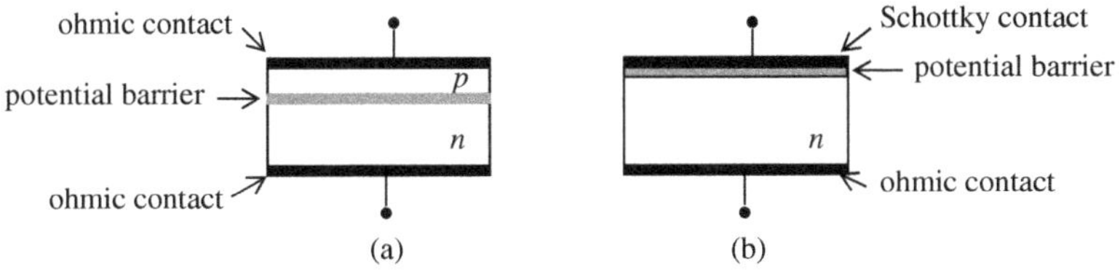

Fig. 3.6. (a) *p–n* junction and (b) Schottky contact based devices in the diode configuration.

For starters, it needs to be emphasized that in the functional devices formation of the *p–n* junction in the way shown for the purpose of illustration of the concept in Fig. 3.3 is not possible. In practice, such devices are formed either by converting by doping part of the *n*-type semiconductor into *p*-type, or *p*-type into *n*-type, or by depositing the *p*-type on *n*-type or *n*-type on *p*-type substrate. Either way the junction formed is parallel to the wafer surface and two terminals are formed on the front and back surfaces of the wafer (Fig. 3.6(a)). The same consideration applies to the geometry of Schottky diode shown in Fig. 3.6(b).

In electronic circuits diodes act primarily as rectifiers converting alternating current (AC) to direct current (DC). This feature makes *p–n* junction diodes indispensable in essentially any type of power supply and/or charger used with phones, computers, TVs and radios as well as with any other stationary or portable electronic equipment. For specialized uses, including diodes performing in high-frequency (microwave) regime, designs of *p–n* junction diodes are adapted to the needs of devices known as Zener diode, varactor diode, and tunnel diode.

In general, selection of semiconductor material used to make a diode depends on its designated applications. For instance, for high-frequency operation materials selected are typically featuring high electron mobility, while diodes designed for high-power/elevated temperature operations are manufactured using wide bandgap semiconductors such as silicon carbide, SiC, or gallium nitride, GaN. In conventional applications of semiconductor diodes, silicon remains the most used semiconductor material.

In terms of diode configuration, the *p–n* junction diodes play dominant role in mainstream electronic applications. The use of Schottky diodes is limited to some compound semiconductors which do not readily lend themselves to the doping processes, and thus, in which *p–n* junction is difficult to form.

3.4.2 *Photonic Diodes*

In addition to electronic applications diode structure dominates the field of photonic devices which convert electric current to light (light emitting diodes), and devices operating in the opposite direction, that is the devices which convert light to electric current (solar cells and photodiodes) (Fig. 3.1). Together, they account for the major component of the field of semiconductor engineering.

Light emitting diodes: Two-terminal semiconductor devices converting electric current into light are known as light emitting diodes (LED). In LED, current injected into the device supplies charge carriers which subsequently spontaneously recombine (spontaneous recombination) and release energy in the form of light in the process referred to as electroluminescence.

In a *p–n* junction diode constructed using indirect bandgap semiconductors such as silicon, the energy resulting from electron–hole recombination is released in the form of the vibrational wave known as phonon which dissipates heat in the solid lattice. In the case recombination occurs in the *p–n* junction diode constructed using direct bandgap semiconductors such as III–V compounds GaAs, AlGaAs, or GaN, resulting energy is released in the form of the photons with energy and corresponding wavelength falling into visible light part of the electromagnetic spectrum.

Figure 3.7 illustrates the simplified fashion operation of the LED. When diode is forward biased, electrons and holes constituting the current recombine within the space-charge region of the junction and release energy is in the form of light (photons).

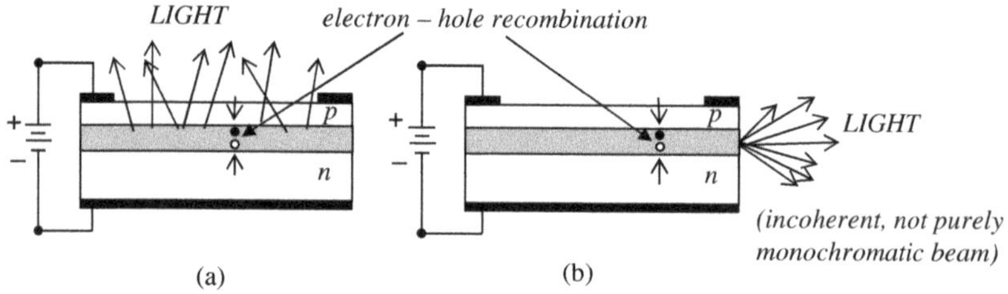

Fig. 3.7. Working principle of the LED: (a) Surface Emitting LED (SELED) and (b) Edge Emitting LED (EELED).

Depending on the construction of the diode, generated light is emitted from the device either in the direction normal to the junction and the surface as illustrated in Fig. 3.7(a) or in the plane of the junction in direction parallel to the surface as shown in Fig. 3.7(b). These two types of light emitting diodes are referred to as the Surface Emitting LED (SELED) or Edge Emitting LED (EELED) respectively. In the former case, thin-film mirrors directing generated light toward top surface are part of the diode structure. While structurally somewhat more complicated, SELED can be more efficiently than EELED coupled with the waveguide which is needed to incorporate light emitting diode into an optical system.

Energy $h\nu$ of photons emitted through spontaneous recombination is equal to the energy gap E_g of semiconductor used to construct LED. Corresponding wavelength λ of emitted light can be calculated using relationship $\lambda\ (\mu m) = 1.24/E\ (eV)$ introduced in Chapter 1. However, because of the way generated light is processed inside the diode, the emerging beam is incoherent and not exactly monochromatic which means that it may contain more than a single wavelength. This is in contrast to the laser diodes considered later in this section.

The wavelengths in the electromagnetic spectrum visible to human eye range from about 390 nm to 700 nm. Using values of E_g of various III–V semiconductors given in Table 2.2 we can see that with the bandgap energies changing from 3.5 eV (GaN) to 1.43 eV (GaAs) and corresponding wavelengths of 360 nm and 870 nm respectively, the III–V compounds, whether binary, or ternary, or quaternary, are able to cover the entire visible spectrum and part of the invisible infrared spectrum.

From the human vision perspective primarily colors, which are the colors that upon proper mixing can produce the entire range of colors are red (λ = 620–750 nm), green (λ = 495–570 nm), and blue (λ = 450–495 nm), or RGB in short. The red LEDs can be engineered using aluminum gallium arsenide, AlGaAs, green LEDs can be implemented using, for instance, aluminum gallium indium phosphide, AlGaInP, while blue LEDs will require use of the wide-bandgap semiconductor such as GaN. As these examples indicate, LED engineering is heavily dependent on the technology of groups III and V elements processed not only into binary, but also into complex ternary and quaternary compounds.

The key feature of the RGB approach is that when placing red, green, and blue LEDs close to each other, and properly adjusting output of each diode, the light white in appearance will result. This is one solution to the engineering of white LEDs (Fig. 3.8(a)) which are at the core of LED

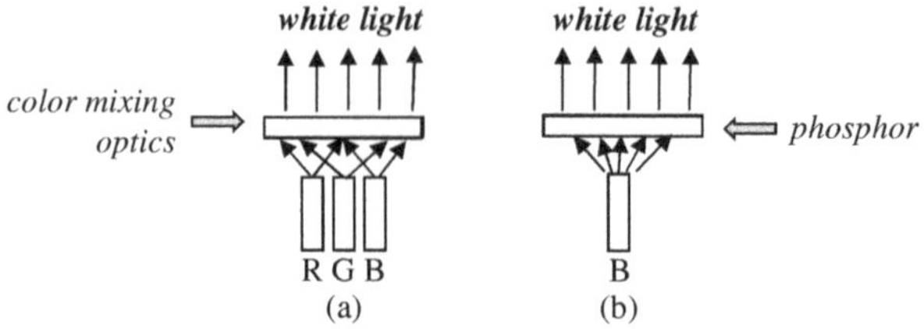

Fig. 3.8. The white light generation by (a) RGB white LED which uses closely spaced red, green, and blue diodes and appropriate color mixing optics and (b) phosphor white LED using blue diode in combination with phosphor.

lighting technology. Another one involves combination of the blue LED and phosphor of desired composition placed in the same bulb envelope (Fig. 3.8(b)). Part of the blue light emitted by the LED and illuminating phosphor initiates through the effect of photoluminescence emission of the yellow light featuring broad spectral power distribution. Remaining blue light, when mixed with yellow light will make light emitted by the diode to be seen as white light.

Either way, with efficiency higher than any other source of white light, incandescent bulbs in particular, LED light bulbs dominate fields of commercial and residential lighting.

Special role in the field of light-emitting diodes is played by organic LEDs (OLED) which use organic semiconductors as an electroluminescing material. Distinct functions OLEDs fulfill in solid-state lighting applications are related to their inherent mechanical flexibility allowing lighting fixtures designs that cannot be implemented using inorganic semiconductors LEDs formed on rigid substrates.

Another common in our daily lives application of light emitting diodes is in display technology. Whether it is in a very large, very bright outdoors displays, billboards, flat-panel TV screens, PC monitors, mobile devices, or destinations signs, LEDs light emission characteristics are being exploited for the purpose of image generation. The LED displays comprise of the active matrix of closely spaced red, green, and blue (RGB) LEDs organized into pixels individually powered by the integrated with each pixel TFT discussed in section 3.5.6. Wherever needed, LED displays can be equipped with a touchscreen features.

Similarly to LED based lighting, also in LED display technology OLEDs play a distinct role in a range of imaging applications from the small cellphones displays to the large TV displays. When using flexible substrates, OLED technology allows flexible displays which bring semiconductor display applications to a different level in terms of convenience and versatility.

Principles of OLED operation are based on the different physical effects than in conventional inorganic LED. In the case of OLEDs it is based on the decay to the ground state (and releasing energy in the process) of excitons formed through interactions between electrons and holes injected from the metal contact and ITO contact into the junction region through the electron and hole transport materials. This contrasts with inorganic LEDs where band-to-band recombination of electron–hole pairs is responsible to the emission of light.

Laser diodes: Light emitted by the conventional LED is incoherent which means that LED does not produce a focused beam of light, and is not exactly monochromatic, or in other words, LED light features a range of wavelengths around dominant intensity line, and thus, range of energies (Fig. 3.7(b)). This characteristic does not hamper usefulness of LEDs in applications discussed above, but is a limitation when there is a need for highly coherent (focused) and monochromatic (single wavelengths) beams of light. To accomplish both these features, conventional LED design needs to be modified into what is known as semiconductor laser, or laser diode (term "laser" is an acronym for "light amplification by stimulated emission of radiation"). The purpose of the modifications in question is to create conditions enforcing in an inorganic semiconductor *p–n* junction stimulated emission in addition to spontaneous emission upon which conventional LEDs operate. The sequence involving spontaneous and stimulated emission initiates a laser action which through optical amplification results in the emission of high intensity, highly coherent, and monochromatic beam of light.

To enforce laser action, the body of the diode needs to include an optical cavity and a very thin layer of intrinsic material between *p*- and *n*-type parts of the junction forming what is known as *p-i-n* diode (Fig. 3.9). It is constructed using the same direct bandgap III–V semiconductor compounds as conventional LEDs, selection of which depends on the desired wavelengths (color) of emitted light.

As Fig. 3.9 shows, a defining feature of the laser diode is the presence of the reflective sidewalls which reflect generated photons causing them to generate more electrons and holes which upon recombination generate more photons. All additionally generated photons are in phase and

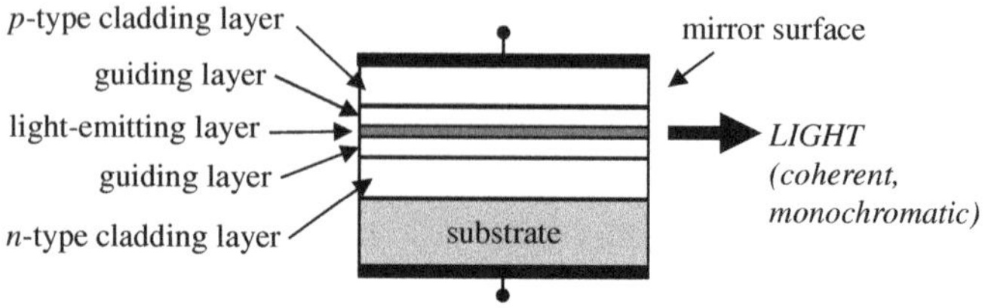

Fig. 3.9. Basic structure of the typical laser diode.

produce coherent, monochromatic beam featuring, unlike conventional LEDs, uniform distribution of power in the beam's cross-section.

The laser diodes are key components in applications too many to list here. The most important uses include fiber optic communications, surgical procedures, optical memories, barcode readers, laser pointers, CD/DVD/Blu-ray disc reading and recording devices to mention just some examples of laser diodes applications.

Solar cells: The term photovoltaics (PV) refers to the technical domain concerned with direct conversion of sunlight into electricity by means of the photovoltaic effect which is an effect underlying operation of semiconductor solar cells. The principles of the solar cell operation is opposite to the operation of the LED where the electric current is converted to light through the electroluminescence process.

In the case of photovoltaic effect, free charge carriers generated as a result of absorption of energy from sunlight remain within the solid and contribute to the current flowing across it. This is contrast to the photoelectric effect in which case electrons generated by absorption of light are emitted to the outside of the solid. Among solids, semiconductors are uniquely suitable for the implementation of photovoltaic effect initiated by the solar cell exposure to sunlight. Hence, the broad term *photovoltaics,* referring often to the related segment of the industry, is essentially a broad reference to the semiconductor solar cells technology.

To make photovoltaic effect work, semiconductor solar cell must be in the form of a diode, typically in the form of two-terminal device with built-in potential barrier associated with *p–n* junction positioned such that it can be readily penetrated by the light illuminating surface of the device (Bisquert, 2017). Figure 3.10 illustrates operation of semiconductor solar cell based on *p–n* junction. As shown in this figure, a distinct feature of the cell is that ohmic contacts to the illuminated surface cover only small portion of the cell surface leaving its remaining portion directly exposed to sunlight. Illuminated surface is covered with a thin layer of material transparent to sunlight which is acting as an anti-reflecting coating. Portion of the sunlight spectrum featuring energy higher than the energy gap E_g of semiconductor causes generation of electron–hole pairs within space charge region of the junction located close to the top surface. Due to the electric field present within the space-charge region, electrons and holes are separated and move by diffusion in opposite directions contributing to the light generated current flowing across the device.

With terminals shorted (Fig. 3.10(a)), a short circuit current I_{sc} results from the photovoltaic effect. When the circuit is open (Fig. 3.10(b)), separation of charge within the cell establishes difference of potential between cell's terminals expressed as an open circuit voltage, V_{oc}. These two key parameters of the solar cell are marked on its output I–V characteristic (Fig. 3.10(c)) shape of which represents parameter known as fill factor, *FF* and which defines performance of the solar cell. In the ideal case $FF = 1$ as depicted by the rectangular area marked in Fig. 3.10(c). In practice, *FF* may vary from about 0.7 to about 0.9 reflecting losses of power related to the quality of the cell.

Parameter most commonly used to define performance of the solar cell is cell's efficiency, η, defined as a ratio of maximum power generated by the cell P_{max} and input power of the solar energy P_{in} ($\eta = P_{max}/P_{in} = I_{sc} V_{oc} FF/P_{in}$).

Efficiency of the solar cell is a function of material from which it is constructed and complexity of its structure, and thus, complexity of the processing steps employed in its manufacture. A universally valid rule is

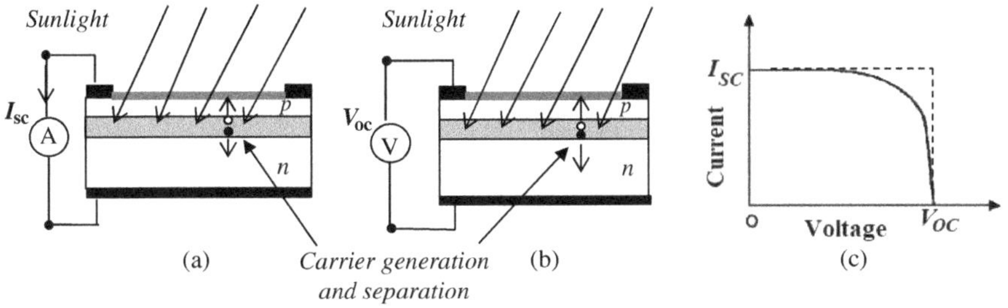

Fig. 3.10. Generation and separation of electrons and holes results in (a) short-circuit current I_{sc}, (b) open-circuit voltage V_{oc}, and (c) output characteristics of a solar cell with short circuit current I_{sc} and open circuit voltage V_{oc} indicated.

that the efficiency of solar cell is proportional to its cost which encompasses cost of the material used and cost of the manufacturing process.

Vast majority of solar cells is manufactured using silicon which is on one hand by far the most common and highly manufacturable semiconductor material, and on the other the energy gap of which sufficiently well matches energy spectrum of the sunlight. Performance of Si solar cells is determined by the cost/quality of the starting material used. Relatively high cost cells manufactured with very thin single-crystal Si wafers may feature efficiency in the 20–25% range, while efficiency of cells manufactured using lower cost multicrystalline Si wafers is around 18–20%. The low-cost commercial thin-film amorphous Si (a-Si) solar cells typically formed on the ITO covered glass feature efficiency around 10%. It needs to be noted, though, that the efficiency of all types of Si solar cells increases with improvements in manufacturing technology and the approximate numbers quoted in this discussion may not represent state-of-the-art at the time this book will be published.

In order to meaningfully increase efficiency of solar cells, multi-junction, multi-material cells known as tandem solar cells must be used with the goal of more efficient use of the solar spectrum by capturing its larger portion. Problem with single material cells is that sunlight's photons featuring energy E lower than bandgap of the cell material ($E < E_g$) do not contribute to the photovoltaic effect at all, while photons with energy significantly exceeding E_g contribute only partially. To alleviate this deficiency, semiconductors featuring different energy gaps need to be stacked up in multi-junction tandem solar cells. The result is a multi-layer structure formed typically using III–V semiconductors of varied composition arranged such that energy gap width increases from the bottom to the top of the stack. This arrangement assures absorption of the broad range of wavelengths featured in the solar spectrum with shortest being absorbed by the top layers, and longer penetrating deeper into the stack where they are absorbed by the narrower bandgap materials. The result is the highest cost class of solar cells, but also the cells featuring the highest efficiencies approaching 50%.

At the other end of the cost-efficiency paradigm are solar cells, manufactured using organic semiconductors. Similarly to OLEDs, the mechanism of light conversions into electricity here is somewhat different than in the case of inorganic semiconductor cells discussed above and also different is efficiency which in the case of organic solar cells is in the range of just a few percent. Despite this, low cost of organic cells combined with flexibility of organic molecules allowing their use in specialized applications where rigid substrates based inorganic cells cannot be used, makes organic photovoltaics a viable technology.

In conclusion of this brief overview of semiconductor solar cells, it needs to be pointed out that the cells from both ends of the cost-efficiency spectrum find useful applications. The choice depends on the type of application which defines how much area can be devoted to the solar panel. If the areas that can be devoted to solar cell panels are counted in square kilometers (e.g. solar farms) then the lower cost, but also lower efficiency, cells could be a solution. If the area is limited, for instance, area of solar panels powering satellites, then high efficiency cells are the only solution, and the cost becomes a secondary issue.

Photodiodes: As the name implies, photodiodes are devices which, similarly to solar cells, convert light into electricity. Also, similarly to solar cells, photodiodes are using *p–n* junction as a potential barrier. The difference lies in photodiodes being designed to respond to specific range of wavelengths (e.g. infrared spectrum) rather than to broad range of wavelengths within the solar light spectrum.

Semiconductor materials used to construct photodiodes are selected based on their energy gap matching wavelength of the light to which photodiode is designed to respond as well as its application. As an example, printed organic photodiodes can be used in wearable applications for health monitoring purposes

3.5 Transistors

The main goal of this section is to introduce and briefly discuss the concept of key devices in semiconductor engineering known as *transistors*. The transistor is a device featuring three terminals and acting as a functional extension of the two-terminal diode by allowing signal amplification and efficient on/off switching. While the concept of transistor was patented long time ago (Lilienfeld, 1930), but it was not until transistor action was realized in 1947 by W. Shockley, J. Bardeen, and W. Brattain and described (Bardeen and Brattain, 1949), that transistors have become drivers of the progress in electronics we are experiencing. Considerations below review principles upon which operation of transistor is based and identify key classes of transistors.

3.5.1 *Transistor Action*

The transistor, or transient varistor (varistor is an electronic element which resistance changes with applied voltage), is a semiconductor device equipped with three terminals (1, 2, and 3 in Fig. 3.11) arranged in such way that the 1–2 input signal in the form of either current or voltage can be used to control 1–3 output current. Since in some cases lower power input signal can be used to control higher power output signal, properly configured transistor is capable of signal amplifying action. Another function efficiently performed by the transistor is signal switching action. Since neither of these fundamental in electronics functions can be efficiently performed by the two-terminal devices, transistors play unique, irreplaceable role in electronics.

As stated earlier, to control the current of any semiconductor device, either number of charge carriers moving across the region of device featured by the constant resistance must be changed, or resistance of such region must be changed. These two concepts are underlying the development of two different classes of transistors. The first is concerned with the bipolar transistor, also known as a bipolar junction transistor, BJT, comprised of two *p–n* junctions where both majority and minority charge carriers contribute to the transistor action justifying its bipolar denomination. The second one includes a class of field-effect transistors (FETs), operation on which is controlled by majority carriers only, and hence, are referred to as unipolar transistors.

3.5.2 *Types of Transistors*

There are many types and sub-types of transistors designed and manufactured to perform specific electronic functions such as signal switching or amplification under the specific operational conditions involving high-power signal, or very high-frequency signal, or both. They all fall,

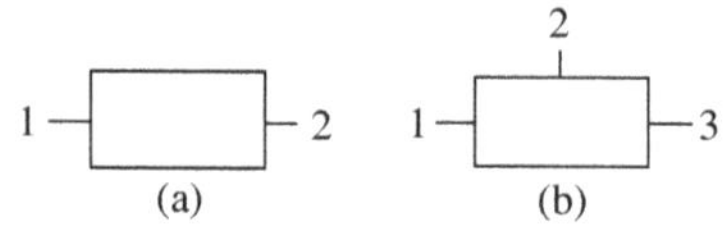

Fig. 3.11. Schematic representation of (a) two-terminal device (diode) where 1–2 voltage controls 1–2 current and (b) three-terminal device (transistor), where 1–2 voltage controls 1–3 current.

however, into one of the two major classes of transistors, bipolar junction transistors, BJTs, and unipolar FETs referred to above.

Bipolar junction transistor (BJT): The bipolar junction transistor is built by adding a second *p–n* junction to the structure shown in Fig. 3.6(a). and by providing additional terminal (ohmic contact) needed to access added junction. The resulting transistor structure is schematically illustrated in Fig. 3.12(a) where three terminals called emitter (E), base (B), and collector (C) replicate terminals 1, 2, and 3 depicted in Fig. 3.11. In the case of BJT in its fundamental version, it means that the current flowing across E-B junction can be used to control current flowing from the emitter to collector across the C-B junction. Depending on application in which BJT is used, most notably applications concerned with signal amplification and signal switching ("on/off" operation), terminals and junctions biasing scheme can be arranged into common-base, or common-emitter, or common-collector configuration. In its very essence, BJT is a device amplifying current, but incorporated into adequately designed circuits it can also be used to amplify voltage or power.

Using common-base configuration as an example, operation of BJT can be explained in the simplified terms as follows. In normal junction transistor's operation mode the E-B junction is forward biased and C-B junction is reverse biased. Under these conditions, large emitter current I_E involving majority carriers (electrons in the case of device in Fig. 3.12(a)) is injected from the *n*-type emitter to the *p*-type base where it becomes a minority carrier current. At the same time current of the reverse biased C-B junction is controlled by the limited in quantity minority carriers. Therefore, by supplying additional minority carriers injected from the

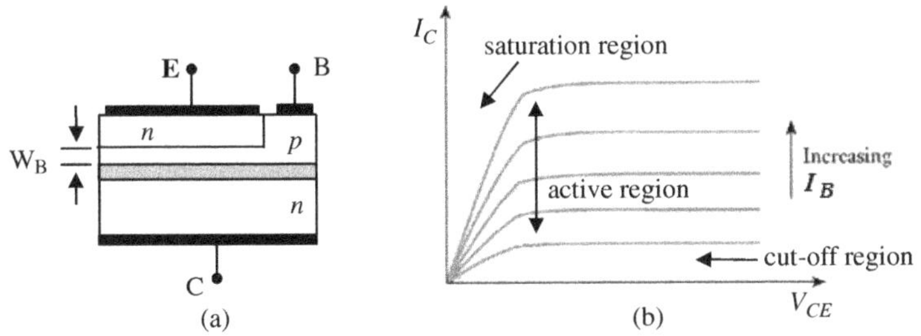

Fig. 3.12. (a) Schematics of BJT and (b) output I–V characteristics in common-emitter (CE) configuration.

emitter, the reverse current of the C-B junction, which is collector's current I_C and at the same time transistor's output current, can be significantly increased. The key to the successful implementation of this sequence is that as many as possible carriers injected from the emitter reach C-B junction. Therefore, it is of critical importance to the operation of BJT that the distance minority carriers must cover between E-B and C-B junctions (base width, W_B, in Fig. 3.12(a)) is as short as technologically possible.

The $I_C - V_{CE}$ plots of the BJT in common-emitter configuration shown in Fig. 3.12(b) can also be also used to illustrate how the switching of electrical signal can be implemented in this class of transistors. The idea is to switch transistor from *fully-OFF* condition (output current I_C as close to zero as possible) to *fully-ON* condition (output current I_C as high as possible) which in terms of output characteristics means changing biasing conditions such that transistor's operation transitions from the cut-off region to the active region (Fig. 3.12(b)).

Shown in Fig. 3.12 is *n–p–n* version of the bipolar transistor (*n*-type emitter, *p*-type base, and *n*-type collector). The concept of the *p–n–p* bipolar transistor can be explained in the same way as above except that roles of the majority carriers and minority carries are reversed. In the case of BJT built into silicon, the *n–p–n* transistors are more often used in practical applications than their *p–n–p* counterparts. The reason is that in the former case higher mobility electrons rather than lower mobility holes are responsible for the charge transfer across the base resulting in the superior performance of the BJT in *n–p–n* configuration.

At this point it needs to be emphasized that in the single *p–n* junction device (diode in Fig. 3.6(a)) no current amplification is possible under any bias conditions. Also, AC signal switching, while theoretically feasible with a junction diode by changing bias voltage from forward to reverse and back to forward, is too slow, too power consuming, and overall inefficient to be considered for any demanding practical application. That is the reason why extending single-junction, two-terminal diode into two-junction, three-terminals bipolar junction transistor expands technology of *p–n* junction-based devices into a wide range of applications in electronic circuits.

By virtue of their characteristics, the bipolar junction transistors are common semiconductor devices in the wide range of electronic applications. They are used primarily in the signal amplification related applications in analog circuits. Because they perform very well in high frequency operation regime, BJTs are commonly used in wireless communication systems. They also find frequent uses as switches in high-power digital circuitry.

Vast majority of commercial bipolar junction transistors are fabricated using silicon. In some cases, germanium is being used for its higher charge carriers mobility as compared to silicon. When a very high frequency operation is needed, bipolar transistors are constructed as Heterojunction Bipolar Transistors, HBTs in short, using high-electron mobility III–V compounds such as gallium arsenide and its derivatives.

Unipolar FET: As pointed out earlier, the second major class of transistors is concerned with unipolar transistors in which current consisting of majority carriers flows in the direction parallel to the potential barrier plane and parallel to the surface. In this case voltage-controlled variations of the conductance of the region adjacent to the surface through which carriers are moving are used to affect flow of carriers, and hence, transistor current. Due to the nature of their operation, transistors of this type are referred to as the field-effect transistors (FETs). Three types of the FETs are distinguished based on the way the space charge region, which expansion is used to control current flow, is induced (Fig. 3.13).

In the case space charge region results from the presence of a *p–n* junction, transistor is called Junction FET, JFET (Fig. 3.13(a)). In the case the same effect is accomplished by means of a Schottky contact, transistor is called Metal-Semiconductor FET, MESFET (Fig. 3.13(b)). The third option is to use ability of MOS capacitor to alter conductance of the near-surface region of semiconductor and to implement Metal-Oxide-Semiconductor variation of the FET known as Metal-Oxide-Semiconductor FET (MOSFET) (Fig. 3.13(c)).

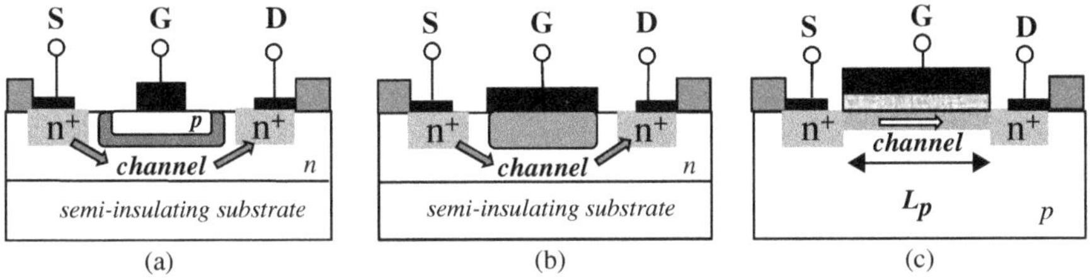

Fig. 3.13. Three types of FETs: (a) Junction FET (JFET), (b) Metal (Schottky contact) FET (MESFET), and (c) Metal–Oxide–Semiconductor FET (MOSFET).

As Fig. 3.13 indicates, all types of FETs are equipped with three terminals called source (S), gate (G), and drain (D) which correspond to terminal 1, 2, and 3 in Fig. 3.11, respectively. According to the scenario considered earlier, the device output current 1–3, which is a drain current I_D, is controlled by voltage 1–2 which is here a gate voltage V_{GB}, in short V_G. To assure adequate performance of the S and D contacts, highly doped regions are commonly formed as parts of the source and drain ohmic contacts.

The basic idea underlying operation of the FETs in Fig. 3.13 is the same and comes to the control over the conductance of the region between source and drain terminals, referred to as a channel, by the field effect inducing voltage applied to the gate terminal. The difference between FETs (a), (b), and (c) in Fig. 3.13 is in the way field effect is implemented to control conductance of the channel.

In the case of JFET in Fig. 3.13(a), expansion of the space-charge region created by the reverse bias of the p^{+}-n junction confines flow of electrons between source and drain to the point where it can be almost completely prevented. In the case of the MESFET (Fig. 3.13(b)), the same task is implemented by changing expansion of the space-charge region associated with the reverse biased Schottky contact. Finally, as considered in more details below, the MOSFET (Fig. 3.13(c)), controls source-drain current using gate voltage to open the channel (inversion layer at the surface of semiconductor), or to close a high-conductance channel (no inversion layer) between source and drain terminals.

In terms of applications, key electronic functions both analog and digital can be performed by essentially any transistor either bipolar or unipolar. However, among transistors introduced above, MOSFET is the most used in the broad and diverse range of applications. The reasons for the lead role of the MOSFETs in state-of-the-art electronics are as follows: (i) MOSFET offers the most efficient performance in terms of the combination of power dissipation and signal delay time in digital circuitry, (ii) its geometry is the easiest to scale down which makes MOSFET and its variations the key building blocks of the majority of high-density integrated circuits, (iii) MOSFET drives evolutionary trends in the field of semiconductor electronics, and thus, in the best way represents state-of-the art in transistor technology, (iv) MOSFET features structure which unlike other types of transistors lends itself to the implementation in thin-film technology, (v) unlike other transistors' designs, MOSFET is compatible with organic semiconductors, flexible electronics, as well as 1D and 2D nanoscale material systems.

It needs to be pointed out that there are some highly specialized applications particularly demanding in terms of speed of operation or power handling, which enforce departure from the basic MOSFET architecture. Overall, however, MOSFET remains the most often used in practice transistor. Therefore, considerations focused on the MOSFET adequately represent technical trends and advancements concerned with the class of semiconductor devices known as transistors.

3.5.3 *Operation of the MOSFET*

MOS Structure: A basis upon which MOSFETS are operating is MOS capacitor. Electrical characteristics of the metal-oxide-semiconductor device are to a significant degree determined by the thickness of the oxide in MOS structure (Fig. 3.14). In the case of ultra-thin oxide, the MOS structure displays diode-like characteristics due to the current tunneling across the oxide. When the oxide is thick enough to prevent current flow across the oxide by tunneling, the MOS structure display characteristics of a capacitor. While not used in electronic circuits as a stand-alone capacitor, the role of the MOS capacitor as a key element of the most important class of transistors needs to be recognized. Basic characteristics of MOS capacitors in configuration compatible with transistor's structure are briefly considered here.

The MOS capacitor is essentially a parallel-plate capacitor with oxide sandwiched between metal contact referred to as a gate and semiconductor (Fig. 3.14(a)). With oxide thick enough, the current flow in the

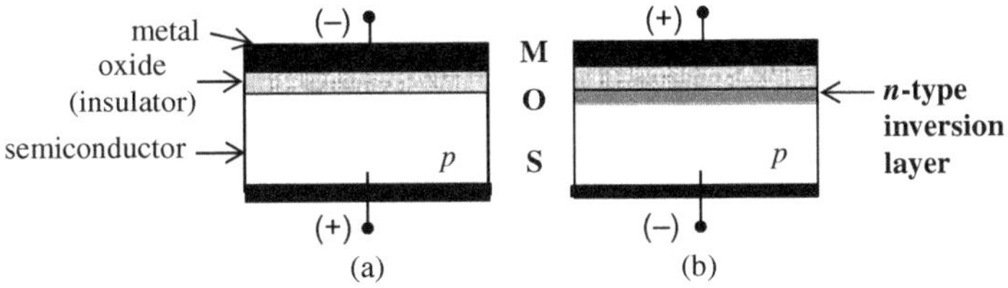

Fig. 3.14. Metal-oxide-semiconductor (MOS) capacitor with (a) negative voltage on the metal gate and (b) positive voltage on the metal gate.

direction normal to the surface of the substrate is prevented regardless of the bias. As a result, such two-terminal structure cannot act as a diode in a way similar to the *p–n* and metal-semiconductor junctions discussed earlier.

Instead of controlling charge flow between metal gate and semiconductor, MOS capacitors are used to implement field effect, considered in Chapter 1, for the purpose of controlling current flow in the direction parallel to the surface by altering conduction of the near-surface region of semiconductor. The desired outcome is accomplished by means of electrostatic interactions between metal gate and semiconductor by means of effects discussed below.

Figure 3.14(a) shows MOS structure with *p*-type semiconductor under the negative voltage applied to the gate. The negative gate potential attracts positive free holes in the semiconductor and causes accumulation of the holes near the oxide-semiconductor interface with concentration of the holes exceeding their concentration in the bulk. When the positive voltage is applied to the gate (Fig. 3.14(b)), positive holes are pushed away from the oxide-semiconductor interface leaving behind thin, near-surface region depleted of free holes, but containing immobile, uncompensated negative acceptor ions. At the low positive gate voltage, the negative ions will prevent penetration of this depletion region by free electrons from the bulk of the semiconductor. At the higher positive gate voltages, however, the inflow of electrons to the surface region depleted of holes will occur and eventually concentration of electrons in the region immediately adjacent to the interface with oxide will exceed concentration of holes. As a result, an inversion region, or in other words a region featuring opposite conductivity type than the substrate, will be formed. In the case considered, *n*-type surface inversion region is formed in the near-surface region of the *p*-type substrate.

As explained in the following section, the ability of MOS gate structure in Fig. 3.14 to invert conductivity type and to create an inversion layer at the semiconductor surface by means of the field effect is a foundation upon which operation of the most important type of transistor, appropriately referred to as metal–oxide–semiconductor field-effect transistor (MOSFET) is based. In addition, MOS capacitors are basis upon which Charge Coupled Devices, CCD, commonly used in cameras as image sensing devices, are constructed (see Section 3.7).

Figure 3.15 shows schematic diagram of the MOSFET, also known as Insulated-Gate FET, or IGFET. Its operation is based on the MOS gate

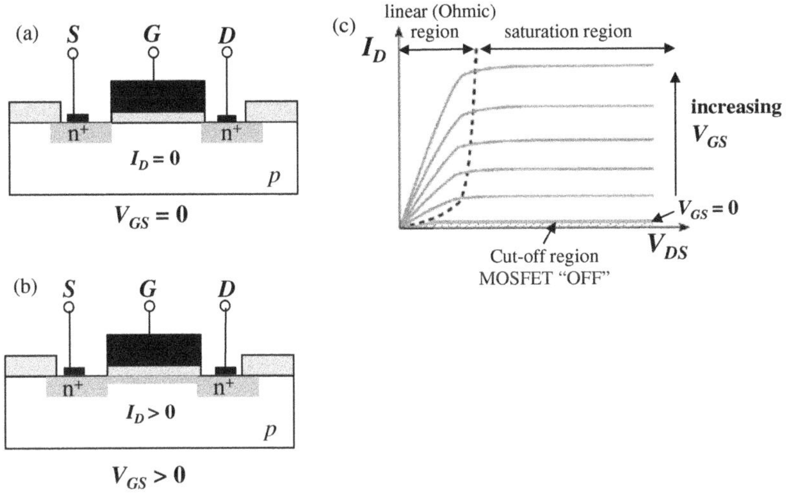

Fig. 3.15. N-channel MOSFET (a) at $V_{GS} = 0$ and with no channel and drain current $I_D = 0$, (b) at $V_{GS} > 0$ and channel created and the drain current $I_D > 0$, and (c) output characteristics of the MOSFET.

capacitor's ability to invert the near-surface region of semiconductor as indicated in Fig. 3.14. With no voltage V_{GS} applied to transistor's gate (Fig. 3.15(a)) there is no inversion layer at the semiconductor surface underneath the gate contact, and thus, there is no channel between source and drain. Transistor is effectively in the "OFF" state ($I_D = 0$) as only negligibly small leakage current is allowed to flow between *S* and *D* terminals.

When the positive voltage exceeding threshold inversion point, or in other words exceeding threshold voltage V_T is applied to the gate, a state of inversion is created at the semiconductor's surface region underneath the gate and the channel between source and drain regions is formed (Fig. 3.15(b)). Under such conditions transistor is turned "ON" and the drain current I_D (MOSFET's output current) flows between *S* and *D* terminals.

The output characteristics of the MOSFET shown in Fig. 3.15(c) differ between various types of MOSFETs depending on the combination of materials (gate contact material, gate oxide material, and semiconductor material) used to construct a transistor. In the case considered here, there is no channel at the gate voltage $V_{GS} = 0$, and hence, output current $I_D = 0$. The gate voltage larger than threshold voltage V_T ($V_{GS} > V_T$) needs to be applied to the gate to create a channel. This type of a MOSFET is called an enhancement mode transistor and is also referred to as a normally "OFF" transistor. A MOSFET in which due to the inherent properties of materials selected to form transistor, channel exists when $V_{GS} = 0$ is called a depletion mode transistor and is also referred to as normally "ON" transistor.

As Fig. 3.15(c) shows, I_D-V_{DS} characteristics at any given V_{GS} feature transition from initially linear regime into saturation regime at higher V_{DS} voltage. The I_D current is saturating with increasing V_{DS} voltage because of changes in the inversion charge density distribution along the channel caused by the voltage drop across the oxide in the direction of the I_D current flow. As a result, gradual thinning of the inversion layer starting at the drain end of the channel occurs. At the V_{DS} voltage at which drain current saturates ($V_{DS} = V_{DS(sat)}$), thickness of the inversion layer at the drain terminal is zero and the channel pinch off occurs. As the V_{DS} continues to increase beyond $V_{DS(sat)}$, a pinch off point at which inversion charge density is equal zero moves toward source terminal which means that channel extends only over the portion of the source-drain distance and transistor current effectively ceases to be controlled by the channel's conductance. The effect of channel length modulation is responsible for the saturation of the I_D-V_{DS} plots at $V_{DS} > V_{DS(sat)}$.

The channel length is one element determining time it takes charge carriers to drift from the source to the drain under the influence of V_{DS}, and thus, defining speed of MOSFET operation. Another one is reduced mobility of charge carriers in the MOSFET's channel. because of (i) limited width of the channel confining moving charge carriers, and (ii) charge carriers moving in the immediate vicinity of semiconductor surface which brings about significant surface scattering. As a result, effective mobility of charge carriers in the MOSFET channel is some five times lower than in the bulk of the same semiconductor. The effect of reduced carrier mobility in the channel varies with gate voltage and increases as the electric field in the channel increases and carriers velocity in the channel approaches velocity saturation limit discussed in Chapter 1. The effect is compounded when the length of the channel gets shorter and at the extreme channel length scaling may eventually cause ballistic transport to take over control over the charge carriers motion in the MOSFET's channel.

The effect of the mobility of charge carriers in the channel affecting performance of the MOSFET calls for appropriate selection of the conductivity type of the channel. In the schematic representation of the

MOSFET in Fig. 3.15 the *n*-type channel (inversion layer) is created in the *p*-type silicon substrate. Such device is referred to as N-MOSFET in contrast to P-MOSFET in which *p*-type channel is induced in the *n*-type substrate. As the MOSFET's operation is controlled entirely by the majority carriers, the former is the structure of choice because higher mobility electrons rather than lower mobility holes are carrying current across the channel resulting in the shorter transfer time and substantially faster operation (switching) of the N-MOSFET as compared to the P-MOSFET featuring the same geometry and fabricated using the same materials.

By comparing the way transistor action is implemented using MOSFET and bipolar junction transistor, as well as output characteristics of these two types of transistors (Fig. 3.15(c) and Fig. 3.12(b) respectively) one can point to the key difference between MOSFET and BJT which is the fact that the former is a voltage (V_{GS}) controlled, while the latter is a current controlled (I_E or I_B depending on transistors configuration) device. The MOSFET features much higher input impedance due to the very high resistance of the oxide sandwiched between metal and semiconductor. Its input current is negligibly small, and hence, MOSFET operates at much lower power levels than bipolar transistors. This feature has important implications when it comes to the choices of transistor configuration the best suited for integrated circuits (see Section 3.6). Additionally, important in this respect is the fact that the MOSFET lends itself to dimensional scaling down much easier than the BJT.

As the discussion in the remaining sections of this chapter will indicate, due to its inherent characteristics and resulting outstanding performance in various electronic applications, the MOSFET represents transistor configuration upon which progress in semiconductor electronics relies on.

3.5.4 *Complementary MOS*

The N-MOSFET in its basic configuration is a transistor capable of satisfactory performance in most digital and analog applications. However, in applications particularly demanding in terms of power consumption and dissipation such as those involving advanced integrated circuits, combining an N-MOSFET and P-MOSFET into a complementary pair known as Complementary MOS (CMOS) in short, solves power management problems, and at the same time creates the most efficient and the most widely used semiconductor cell. The idea behind pairing N- and P-MOSFETs is that with one transistor of the pair being always kept in the “OFF” state, the CMOS cell draws power only during the very fast switching between “ON” and “OFF” states. This is a situation which cannot be attained with a single N-MOSFET which prior to broad implementation of CMOS technology was a transistor of choice in common digital circuits applications (Baker, 2019).

Figure 3.16(a) shows schematic diagram of the N-MOS and P-MOS transistors which are combined into complementary pair (Fig. 3.16(b)). The CMOS cell formed in the *p*-type substrate wafer in which *n*-well needs to be created to allow formation of the P-MOS transistor. To prevent any parasitic electrical interaction between transistors in the CMOS cell, known as CMOS latch-up, an isolation trench filled with oxide is formed between two transistors. Overall, with very low power consumption and dissipation, very low energy needed to perform switching, as well as very low standby current, the CMOS is the most effective semiconductor cell in the implementation of digital functions.

In electronic circuits CMOS cells are most often configured as inverters. In the CMOS inverter configuration (Fig. 3.16(b)) the source of the *N*-channel transistor is connected to the drain of the *P*-channel transistor, and the gates are connected to each other. In such case the output (drain

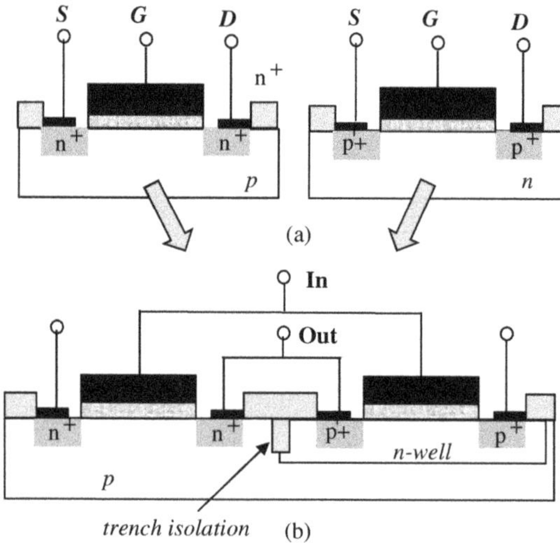

Fig. 3.16. (a) NMOSFET and PMOSFET integrated into (b) Complementary MOS (CMOS) cell in inverter bias configuration.

of the *P*-channel transistor) level of inverter is high whenever its input (gate) level is low, and the other way around. Due to its inherent characteristics CMOS inverter is a basic building block of most digital circuits both logic and memory. Essentially all digital circuits such as microprocessors, microcontrollers, several types of memories, and many others are fabricated employing CMOS technology.

While digital (computational) applications dominate uses of CMOS class of devices, there are also several analog circuits such as operational amplifiers, multiplexers, data converters, as well as transceivers for many types of communication systems that are based on CMOS technology. Due to its versatility, CMOS cells are also commonly used in mixed-signal (analog and digital) applications.

Somewhat special analog application involves use of CMOS technology to fabricate image sensors. CMOS image sensors are an alternative to CCD (Charge Coupled Device) design of image sensors converting optical image into electric signal. The choice between CCD and CMOS image sensors depends on specific application (see Section 3.7).

3.5.5 *Evolution of the MOSFET*

Three elements define, and will continue defining, broadly understood progress in MOSFET technology. First is concerned with dimensional considerations related to transistor's geometry. As it turns out, most of the issues related to MOSFET's geometry are rooted in the process of gate scaling considered below. Second is related to the selection of materials including semiconductors, dielectrics, and conductors which are used to manufacture MOS transistors. The third one is concerned with transistor's architecture which defines shape and configuration of its key parts which evolve as the transistor's performance demands grow.

The elements defined above are considered in the following discussion in terms of the developments regarding transistors in logic applications. As will be apparent from these considerations, the emphasis on each of them varied at the different stages of transistor technology evolution.

Gate scaling: An important geometrical feature of the MOSFET is a distance electrons must cover between source and drain which is denoted in Fig. 3.13(c), and referred to in the discussion above, as channel length, *L*. Schematics of the MOSFET (e.g. in Fig. 3.16), commonly adopted to explain principles of its operation imply that the length of the channel *L* and the length of gate contact L_G formed on top of the oxide are essentially the same. In practical devices, however, due to the nature of the MOSFET manufacturing process, physical length of the channel may be slightly shorter than the length of the gate. The extend of the difference between L and L_G is likely to be different depending on the manufacturing sequence. The gate length L_G represents minimum feature size characteristic of any given transistor technology and as such is used to distinguish various classes of the MOSFETs from the point of view of their operational capabilities.

In the light of the above it is a commonly adopted practice to refer to the gate length as a parameter defining scaling, and hence, performance of the transistor. Shorter gate length results in the faster transfer of carriers from source to drain and leads to improved performance of the MOSFET. Therefore, gate scaling is a driving force behind the progress in advanced transistor technology both in terms of device design and device manufacturing processes. Since the beginning of MOSFET technology to recent technology generations, gate length underwent reduction by three orders of magnitude from 10 μm to the 10 nm and below.

The push toward shorter gate results in the improved transistor's performance. Specifically, if the gate is scaled by the scaling factor k, then the product power × delay time reflecting power handling capability of transistor and its speed of operation is reduced by roughly k^3. Moreover, in the process of scaling, physical size of the entire transistor is decreasing which allows more transistors to be packed per unit area of semiconductor chip.

The gate scaling scenario is relevant only in cases where transistor's highest possible speed of operation and small size are the lead considerations. On the other hand, there are transistors applications, for instance, in power circuits or in thin-film technology, where neither speed of operation, nor ultra-small size of transistor are the key progress defining considerations. Furthermore, gate scaling brings about manufacturing challenges and associated skyrocketing cost of transistor fabrication. As a result, depending on how any given class of transistors is used, manufacturing cost rather than physical limitations may decide on whether further gate scaling is pursued or not. In the cases where the performance of transistor featuring longer physical gate length can be made equivalent due to the improved technology to the performance of the fictitious device featuring shorter gate-length, then the equivalent gate length (EGL) can be used to describe improvements in transistor performance. Either way, progress toward transistor performance equivalent to the performance of transistor sub-nm gate length continues and is not solely dependent upon gate scaling, but also on innovative solutions regarding materials and transistor architecture.

Materials: The discussion regarding materials comprising a MOSFET needs to be concerned with semiconductors forming transistor's channel, dielectric material acting as gate dielectric, conductors forming gate contact as well as ohmic source and drain contacts and heat handling capabilities of the entire structure.

Regarding semiconductors, selection of material is determined by the application of transistor. In most of the mainstream electronic applications silicon remains a material of choice. In the case of transistors designed for high-power, high-temperature uses, wide-bandgap silicon carbide, SiC, is used in the case of the MOSFET, while excellent characteristics of gallium nitride, GaN, are exploited using other than MOSFET configurations of FETs not discussed in this *Guide*. When the speed of transistor operation is at stake, the focus is on the material that assures high electron mobility in the MOSFET's channel. For that purpose, strain can be built into silicon channel (Fig. 2.19), or high electron mobility compound semiconductor material, often in nanoscale configuration, can be integrated with silicon substrate in the channel region.

Requirements concerned with selection of the gate dielectric are defined by the need for the MOS capacitor to provide strong enough capacitive coupling between gate contact and semiconductor to assure inversion of semiconductor's near-surface region and creation of the channel at the as low as possible gate voltage. The needs in this regard can be explained by considering capacitance C of the parallel plate capacitor formed by the MOS structure $C= \varepsilon_0 k\, A/x_{ox}$ where ε_0 is a permittivity of vacuum, k is dielectric constant of the gate oxide sandwiched between metal and semiconductor, A is the area of the gate contact, and x_{ox} is thickness of the oxide.

Due to the scaling of the gate length (L_G), and hence reduced gate area A, the need to maintain sufficient capacitance of the MOS gate stack was met by the gradual decrease of the thickness of SiO_2 gate oxide (x_{ox}). Obviously, such scaling is effective only up to the point where gate oxide is no longer fulfilling its insulating function because of the excessive tunneling current flowing across it. At this point the only way to maintain required density of the gate capacitance at the reduced gate length is to increase dielectric constant *k* of the gate dielectric which for all practical purposes means using as a gate insulator material other than SiO_2 featuring $k = 3.9$ (see Section 2.9.2). As an example, with the gate dielectric featuring *k* five times higher than 3.9, capacitive coupling of the MOS gate which would require let say 0.5 nm thick SiO_2 (Equivalent Oxide Thickness, EOT), can be accomplished at the physical thickness of the gate oxide increased to 2.5. At this thickness of the gate oxide probability of electrons tunneling across it would be reduced significantly (Ruzyllo, 1980).

In advanced silicon-based MOS/CMOS technology choices regarding gate dielectric depend on the gate length (technology node). For the gate length of 45 nm and below, high-*k* dielectrics are used as gate insulators. For all other technology nodes (65 nm and above) silicon dioxide SiO_2 is effectively performing the role of the gate oxide.

The choice between high-*k* and SiO_2 gate dielectric is driving selection of the conductors used as gate contacts. Wherever SiO_2 is used as a gate dielectric, heavily doped polycrystalline silicon (poly-Si), often capped with a layer of silicide (see Section 2.10) is a gate contact of choice. In the case where high-*k* dielectric is playing the role of gate dielectric, metals featuring adequate work function and chemically resistant are used as gate contact materials. The need for metals, as opposed to poly-Si in this last case, is due to the interactions between high-*k* dielectrics and poly-Si leading to the undesired capacitance decreasing effect known as gate depletion effect. The class of MOSFETs incorporating high-*k* gate dielectric and metal gate represent technology known as high-*k*, metal gate (HKMG).

Transistor architecture: In addition to the gate scaling and performance enhancing changes in materials comprising transistor, modifications of the transistor architecture provide yet another avenue toward improvements of its characteristics.

As it is understood here, the term transistor *architecture* can be used interchangeably with the term transistor design used in reference to the way key elements of transistor structure are sized, shaped, and positioned with respect to the substrate, and to each other. The demands in this regard vary depending on application any given type of transistor is designed for. An attempt to consider even some of them would be well beyond the scope of this discussion. Therefore, a brief overview of the performance driven evolution of transistor architecture is limited here to the case of ultra-small, high-speed, low-power MOSFETs commonly used in CMOS configuration and is focused on the transition of transistor architecture from planar to vertical.

Figure 3.17(a) shows conventional planar MOSFET. The characteristic of the planar configuration is that the gate voltage can control conductance of the channel, and hence drain current I_D, from one side only. This means that any scaling of the gate length is bringing about undesired reduction of the gate capacitance in agreement with the discussion of the gate capacitance challenges considered earlier in this section.

A solution to the constraints imposed by the planar architecture is a configuration in which transistor's channel is positioned vertically on its side assuming fin-like geometry (Fig. 3.17(b)). To make it into functional transistor, a fin-shaped MOSFET, or FinFET, is fitted with drain and source regions and the metal-dielectric gate surrounding vertical channel on three sides. As a result, the area of the gate increases significantly while the area transistor occupies on the wafer surface is decreased. Through the increased gate area, capacitive coupling between gate and the channel increases, improving control over the channel conductance, and thus, control over the drain current I_D. The result is improved transistor's electrostatics, or in other words, improved overall ability to control channel conductance using low gate voltage V_G.

In the case shown in Fig. 3.17(b) gate surrounds channel on three sides and is referred to as a tri-gate structure. Even better performance can be obtained with gate-all-around, GAA, configuration which represents an extension of the FinFET concept (Fig. 3.17(c)) in which channel does not originate from the substrate, but is processed separately. Added advantage of FinFET configuration is that it can be converted into multi-gate

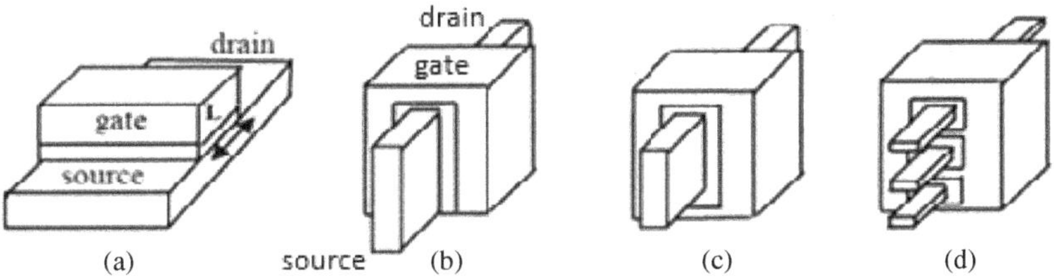

Fig. 3.17. Illustration of changes in MOSFET's architecture needed to maintain adequate gate capacitance at reduced geometry, (a) conventional planar MOSFET, (b) transformed into vertical configuration known as FinFET, (c) FinFET with gate-all-around (GAA) and (d) nanosheet MOSFET.

FET (MuGFET) structure in which transistor's gate is broken into several gates further improving current driving characteristics of the transistor as is in the case of nanosheet FinFET shown schematically in Fig. 3.17(d).

In addition to FinFET, there are other variations of the MOSFET architecture in which vertical configuration is implemented differently. As an example, vertical MOSFET configuration is implemented in the vertical slit FET (VeSFET) in short, (Maly, 1987) which includes two vertically configured gates with a very narrow slit in between them acting as a channel. As simulations indicate, in switching applications transistor in this configuration features advantageous ratio of *on* to *off* current (Pfitzner, 2011). Also, the effect of tunneling can be exploited in transistors featuring ultra-thin gate oxide in yet another configuration (Ruzyllo, 1980; Majkusiak, 1998).

Alternative solutions: Solutions concerned with materials and architecture modifications of the MOSFET discussed above, may not be enough to meet future needs with regard to the performance of transistors needed to carry out increasingly complex computational functions. Therefore, alternative solutions departing from the mainstream approaches to transistors' performance improvements are pursued.

Logic circuits performing computational operations, function based on the binary system needed to code and process information which basically comes to switching the system from one state to another. In electronic devices this function is efficiently carried out by transistors which turn flow of electrons (current) *ON* and *OFF*. The problem is that scattering of electrons moving through the solid accounts for the significant signal losses, and thus, electrons as the carriers of electric charge may not be able to support long-term needs of information processing systems.

One of the solutions being explored is concerned with transistors in which magnetic rather than electric field controls device operation and exploits the fact that in addition to electric charge, electron also features an angular momentum known as electron spin. The spin of the electron can be directed by the magnetic field either up or down. In the sense then, it represents an inherently binary system which spintronics attempts to use in a magnetically sensitive transistor, called a spin transistor, to perform logic functions (Bandyopadhyay and Cahey, 2016).

In another solution, electron as an information carrier is replaced by the photon, representing quantum of light (recall references to photon earlier in this book). Photon is more efficient than electron information carrier as it can cover distances in the waveguides with very little losses, which contrasts with electron moving in semiconductor materials and metals with significant losses. With short wavelength laser diodes and detectors available, and optical waveguides technology being well developed, a high-performance optical transistor is needed to turn light *ON* and *OFF*.

Venturing beyond the realm of the binary system and conventional computers, attention calls a steady progress in quantum computing which is based on the quantum phenomena reaching deep into the inner-atomic properties of solids. The device used to carry out quantum computing operations is referred to as quantum transistor in spite of the fact that its operation is based on entirely different principles than the operation of the conventional MOSFET and as such are not discussed in this book. What these two drastically different in terms of principles of operation transistors have in common, is the fact that both are constructed using primarily semiconductor materials and fabricated using methods which rely heavily on semiconductor nanotechnology.

3.5.6 *Thin-Film Transistor*

A TFT shown in Fig. 3.18 is a MOSFET fabricated using thin-film technology (Machlin, 2010) rather than conventional technology which forms MOSFETs on the bulk wafers.

Unlike in the bulk MOSFET, where channel is formed in the single-crystal semiconductor, the channel in TFTs is most commonly formed using thin-film non-crystalline, amorphous semiconductor. By definition then, TFT features inferior to conventional MOSFET electronic properties

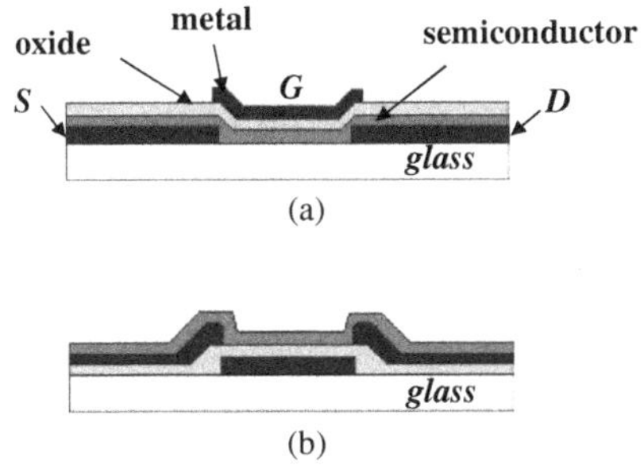

Fig. 3.18. TFT in two different configurations: (a) top gate and (b) bottom gate.

because of the much higher electron mobility in the single-crystal semiconductor as compared to the amorphous semiconductor.

In spite of their performance limiting features, TFTs are among the most important semiconductor devices because of the role they play in flat panel display technology in particular. Whether it is a Liquid Crystal Display (LCD) or emissive display based on Light Emitting Diodes (LED), the best resolution, highest contrast, and significantly improved addressability is achieved when each pixel in individually powered up by the transistor integrated into the pixel's structure. Displays incorporating TFTs are known as Active Matrix displays and offer the best rendering of images and colors.

Figure 3.18 shows schematics of the TFT in two different configurations. The top-gate configuration (Fig. 3.18(a)) reflects the structure of the bulk MOSFET (Fig. 3.15(a)) except that the source and drain and formed using thin-film metal pads (molybdenum, aluminum, and others) and semiconductor in which channel is created by the gate voltage is a thin film semiconductor most commonly amorphous silicon. In some cases, transparent to light wide-bandgap semiconductors such as, for instance, zinc oxide, ZnO, are used instead. A glass substrate shown in Fig. 3.18 is the most common substrate in TFT technology. Depending on application, glass as a substrate can be replaced by other insulating materials both rigid and flexible (e.g. foil). Common in semiconductor device engineering insulators such silicon dioxide, SiO_2, silicon nitride, Si_3N_4, or aluminum oxide, Al_2O_3 are used as gate dielectrics in TFTs while metal gate contacts are processed using metals such as molybdenum, titanium and others depending on the process. To meet the needs of specific application, a TFT can be implemented in the bottom-gate configuration (Fig. 3.18(b)) which operates in the same way as top-gate device.

The nature of thin film technology makes TFTs compatible with the displays manufacturing schemes. As a result, pixels controlling TFTs are an integral part of the display structure. In some types of displays it is essential that TFTs incorporated into the display are transparent to light. Using transparent to light wide-bandgap semiconductors and transparent conductive oxides as contacts, transparency of TFTs can be readily accomplished (which contrasts with inherently opaque to visible light bulk MOSFETs). Furthermore, TFTs provide convenient solution when there is a need for transistors in flexible, wearable electronic devices and circuits. Special case involves the use of TFT structure in the implementation of transistors using organic semiconductors. Organic TFTs (OTFTs) are key components constituting an important class of emissive displays based on the organic light emitting diodes (OLEDs).

3.6 Integrated Circuits

Until now, discussion of semiconductor devices in this chapter was concerned with discrete devices which are single transistors or diodes mounted each in the individual package and used as self-contained elements in electronic circuits. It is a common procedure to construct electronic circuits by attaching, using solder joints, discrete elements to the printed board (Fig. 3.19(a)). Inherently, such circuits are bulky, and relatively expensive with respect to the electronic functions they perform. Most importantly, however, they suffer from reliability problems resulting mainly from the failures of solder joints, and mechanical/electrical connections involved with the use of the circuit formed on the printed boards. Furthermore, printed board technology based solely on discrete transistors for obvious reasons does not allow complex logic circuits involving millions and billions of transistors.

To assembly such circuit, its integration into a self-contained very small volume, fully functional element known as monolithic integrated circuit (IC) is required (Fig. 3.19(b)). The IC technology allows integration of billions of transistors into tiny volume of semiconductor material, such as silicon, for the purpose of formation of electronic circuits designed

(a) (b)

Fig. 3.19. Electronic circuits implemented using (a) discrete elements forming a circuit mounted on the printed circuit board (PCB) and (b) circuit integrated into semiconductor chip representing monolithic integrated circuit technology.

to perform complex electronic functions, for instance, computational. The outcome of the monolithic circuit integration is radically improved performance and reliability at the reduced size and cost of electronic function it performs as compared to the circuits composed of discrete elements (Runyan and Bean, 1990).

The basic idea behind the monolithic integrated circuit technology is to fabricate a complete electronic circuit on a small piece of semiconductor referred to as a chip (Fig. 3.19(b)). In other words, not only transistors comprising a circuit, but also conducting lines interconnecting transistors in the circuit are processed into a tiny piece of, for instance, silicon which is then encapsulated into hermetically sealed package equipped with input/output pins, and ready for installation in the larger electronic circuit. Alternative to monolithic ICs are hybrid ICs where the term "hybrid" refers to the divers technologies that are used to complete the circuit and may include thick-film, thin-film components, as well as monolithic ICs.

Advanced integrated circuits represent the highest level of technical complexity in exploitation of the outstanding features of semiconductor materials and as such act as the technology drivers in the field of semiconductor engineering. An overview of integrated circuits in this section is meant to act as an interface between basic problems of semiconductor materials and devices considered earlier, and the later discussion related to the semiconductor process technology.

Figure 3.20 presents major classes of digital, analog (linear), and mixed signal (digital and analog functions combined on the same chip) integrated circuits distinguished based on the function they are constructed to perform.

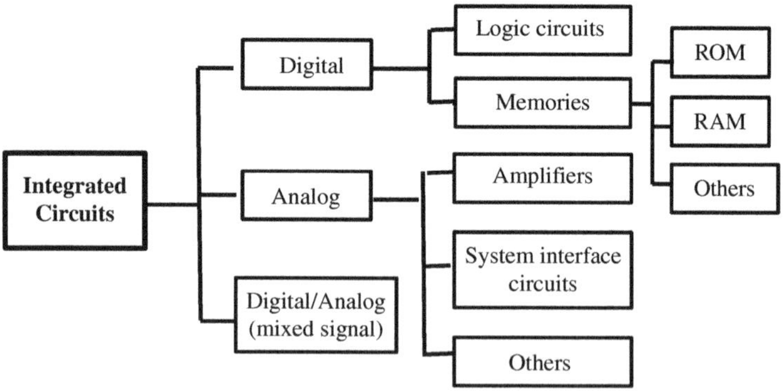

Fig. 3.20. Basic classes of semiconductor integrated circuits (ICs).

Specific layout of the circuit and the choice of materials used in its manufacture are determined by such function. From this point of view, two classes of ICs are represented by the application specific integrated circuits (ASICs) and general-purpose ICs with names clearly identifying the nature of each class. In terms of the circuit's fundamentals, the choice is most often between MOSFET/CMOS and bipolar transistor-based circuit designs. The former are particularly well suited for digital IC applications both in logic and memory ICs, while the latter are typically selected to perform analog functions. In the case of mixed signal ICs, CMOS and bipolar circuits are formed on the same chip.

Considering applications, logic circuits are best represented by microprocessors and microcontrollers with the former performing function of a central processing unit (CPU) in any computer. Memory circuits are designed to store information, temporarily or permanently, and act in support of microprocessors in computers. They fall into two major categories of Random Access Memory (RAM) and read only memory (ROM).

As far as analog ICs are concerned (Fig. 3.20), their uses vary broadly from simple circuits such as operational amplifiers, power supplies and regulators, oscillators and filters to the structurally complex circuits, usually ASIC, designed to carry out analog functions at the frequencies up to the range of thousands GHz. In the past microwave ICs relied primarily on high-electron mobility semiconductors such as gallium arsenide (GaAs). Nowadays silicon (Si), silicon germanium (SiGe), and, increasingly, gallium nitride (GaN) capable of operating at high voltages and power levels at microwave frequencies are also used to fabricate monolithic microwave ICs.

The third class of ICs listed in Fig. 3.20 is concerned with mixed-signal integrated circuits which combine digital and analog functions on the same chip and constitute essential parts of almost any application specific electronic system designed to perform functions that go beyond the rudimentary ones.

In the case of MOSFET/CMOS based digital integrated circuits, logic in particular (Fig. 3.20), scaling of the gate length is a key to the improvement of transistor performance, and thus, is a factor marking progress in integrated circuit engineering. However, gate length cannot be scaled down without adjustments being made to the other parts of the transistor structure. Not following rules in this regard, known as scaling rules, will prevent transistor in the integrated circuit from performing properly. In general, the scaling rules require that the scaling of the gate length is followed by scaling of other geometrical features of the MOSFET both lateral (width of the gate contact), and vertical (depth of the source and drain regions).

The result is that transistors comprising an IC are getting smaller and thus, can be more densely packed on the surface of the chip. In semiconductor engineering the number of transistors per chip is defined by the Moore's Law according to which number of transistors per chip was doubled roughly every two years. General trend regarding increase of the number of transistors per mm^2 of the chip area over the last five decades is illustrated in Fig. 3.21.

Noted progress in IC technology is affected by additional challenges the increase in chip density is facing. First of them is concerned with management of the heat produced by densely packed logic IC chips performing computational operations. The other results from the increasing complexity of the network of electrically conductive lines connecting trillions of transistors into a functional electronic circuit.

The issue of heat management itself presents a challenge because as inherently as the concept of increased integration favors the speed of operation of electronic circuits, it imposes limits on the power handling capabilities of the ICs (Liang *et al.*, 2017). The very small size of devices in the ICs results in a heavy local accumulation of heat generated by the current carried by devices, which in extreme cases may lead to its destruction (see discussion related to Fig. 1.14). Even if not so, the dissipated heat raises the temperature of the adjacent elements closely spaced in the silicon chip affecting their operation.

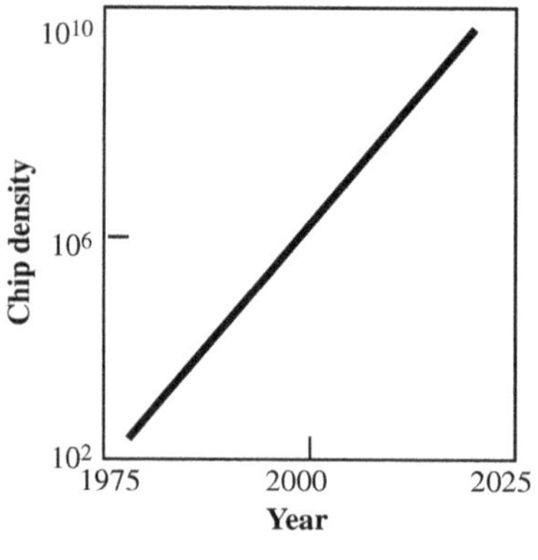

Fig. 3.21. Variation of the chip density (number of transistors per mm^2 of chip area) in digital IC technology over the years.

A key component of any integrated circuit other than transistors themselves is a network of conducting lines formed on the chip surface in the form of thin-film metals needed to electrically connect transistors comprising a circuit. As the geometry of transistors in the circuit is scaled down, the interconnect lines should also be subject to scaling to limit the area of the chip devoted to interconnects and to save chip real estate for further expansion (Fig. 3.22).

The problem is that in contrast to transistors scaling, which results in significant improvements of the performance of the circuit, reduction of the geometry of interconnect lines in the integrated circuits has an adverse effect of their performance. This is because scaling down geometry of the metal film in nanometer-scale brings about increase of its resistance, and thus, has an adverse effect on the current flow. Assuming that with transistor's scaling the line's width W and thickness d (Fig. 3.22(a)) is reduced by the scaling factor k (W/k and d/k, Fig. 3.22(b)), the density of current J in the line ($J = I/Wd$ in Fig. 3.22), is increased by k. In the nanometer-scale geometry of the line such scaling brings current density to the level at which generation of excessive heat associated with the flow of very high density current results in the destruction of the line.

As the reasoning above indicates, in contrast to transistors comprising a circuit, geometry of the interconnect lines cannot be scaled down. Instead, to accommodate the needs of the continued growth of circuit complexity without devoting significant part of the chip surface to the interconnect network, the conducting lines in the circuit instead of being formed in the single level are stacked up in the multilevel metallization configuration. Figure 3.23 illustrates the concept in schematic fashion.

What multilevel metallization scheme brings about is a need for an effective insulation of the adjacent metal lines needed to prevent highly undesired *crosstalk* between metal lines closely spaced in the stack. Such *crosstalk* could have a fatal effect on the operation of the circuit particularly at higher frequencies. To effectively de-couple, or in other words to

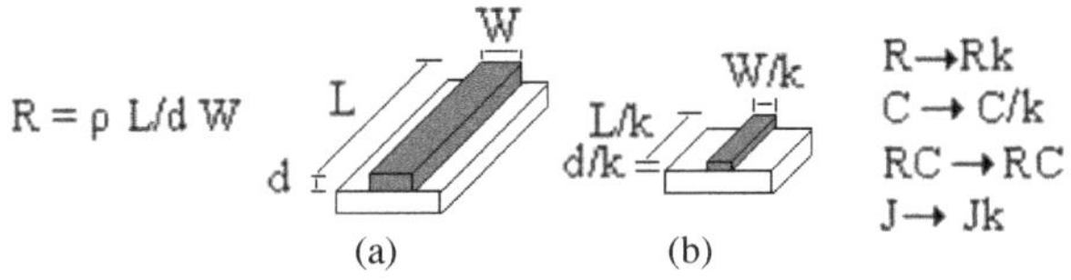

Fig. 3.22. Scaling of the interconnect line geometry by k is causing increase of current density J by k.

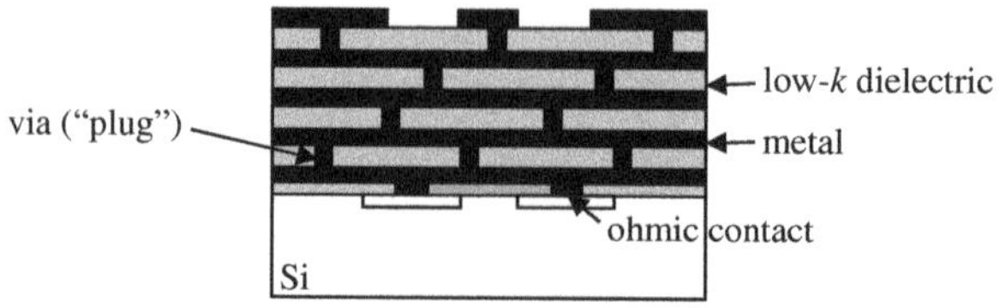

Fig. 3.23. Schematic representation of multilevel metallization (multilevel interconnect) scheme employed in high-density integrated circuits.

minimize capacitive coupling between closely stacked interconnect lines, insulators used must feature as low dielectric constant k as possible (Fig. 3.23). As discussion of low-k dielectrics in Section 2.9.2 indicates, low-k dielectrics are those featuring k between 1 (dielectric constant of air) and 3.9 (dielectric constant of silicon dioxide SiO_2). Among the ways to lower k and to bring it as close to 1 as possible, derivatives of SiO_2 modified toward lower k by adding porosity (nanoglass) and selected organic compounds are used.

A conclusion from the brief overview of selected aspects of integrated circuits engineering is that the scaling of geometry of devices comprising an IC must be accompanied by other innovations to assure better performing circuits for the IC technology generations to come.

3.7 Image Displaying and Image Sensing Devices

Among semiconductor-based consumer products important role is played by the image displaying (displays) and image sensing (image sensors) devices. The function of the former is to process electrical signal into image appearing on the display. The latter, play a lead role in digital systems designed to capture and process images by converting light signal into electrical signal (Tang and Saeedkia, 2016).

The displays related applications of semiconductor materials were considered earlier in this chapter and will be additionally discussed later in this volume. The image sensing devices, often referred to as imaging devices, are designed to capture light representing image and to convert it into electrical signal in a way which can distinguish between different intensities within the light beam illuminating the device. In short, semiconductor image sensors are an image sensing and recording part of any imaging devices such as digital still camera (DSC), digital video camera (DVC), mobile devices, medical, surveillance, scientific, and

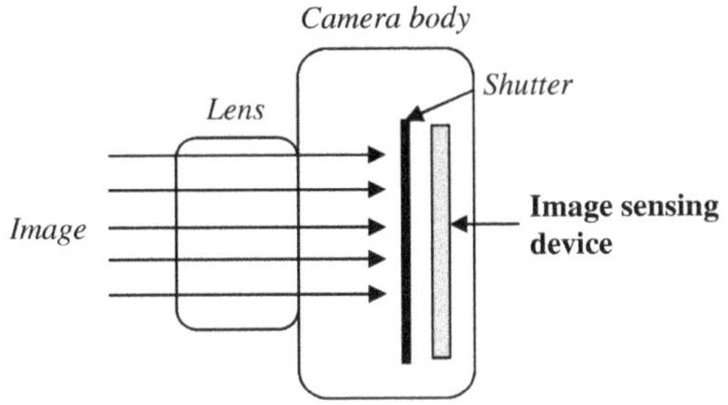

Fig. 3.24. The image sensing device inside the digital still camera.

broadcast instrumentation. Figure 3.24 schematically illustrates the role of the images sensor using single-lens reflex digital still camera as an example.

Due to their physical properties, semiconductor materials are uniquely suited for the manufacture of imaging devices. As a result, imaging devices constitute a self-contained, viable segment of commercial semiconductor technology. Two the most important types of semiconductor imaging devices are both based on the physical properties of the MOS structure. First involves CCD (Charge Coupled Devices) image sensors while the other CMOS (Complementary MOS) image sensors. In contrast to solar cells where incoming light generates free charge carriers forming electrical current, MOS based imaging devices exploit sensitivity of charge distribution in the space charge region of MOS capacitor/gate to the intensity of incoming light which is then converted into the electrical current. Due to these characteristics, CCD and CMOS imaging devices do not need to use semiconductor material featuring specific energy gap and type of the energy gap (direct, indirect) and thus, are manufactured using the most common semiconductor which is silicon and take advantage of the fact that silicon based MOS device technology is well established.

Depending on specific application defined by the needs regarding resolution of the image capturing, power consumption, cost, and size, either CCD, or CMOS image sensors are used. The former are generally used in high-quality imaging devices such as broadcast quality video cameras where small size, cost, and power consumption are not primary concerns. The latter are commonly used in portable/mobile battery powered consumer products, e.g. still photography cameras (Fig. 3.24) or smartphones in which size, cost, and power consumption are dominant concerns. Furthermore, CMOS sensors allow low-light imaging compatible with security and surveillance cameras potentially operating at the varying ambient temperatures.

3.8 Micro–Electro–Mechanical Systems (MEMS) and Sensors

A distinct class of semiconductor devices is represented by MEMS, also referred to as nano–electro–mechanical systems (NEMS) depending on the size of device features. The electro-mechanical semiconductor devices, representing class of devices depicted in Fig. 3.1, were conceived as a way to exploit excellent mechanical properties of silicon which allows literally endless possibilities for integration of electronic and mechanical functions within a single material system. (Liu, 2011).

Semiconductor based sensors comprise an important group within the family of semiconductor devices (Ren and Pearton, 2017). In the case of sensors, sensitivity of semiconductor material and device characteristics to the external influences such as, for instance, temperature and chemical composition of the ambient, make semiconductors uniquely suitable for the broadly understood sensing applications. A significant number of semiconductor sensors exploit operational characteristics of the micro-electro-mechanical systems which justifies reviewing both within the same section.

MEMS/NEMS devices: Micro–Electro–Mechanical System (MEMS) and Nano–Electro–Mechanical System (NEMS) devices integrate mechanical and electrical functions using somewhat modified, but otherwise standard semiconductor device manufacturing technology. Such functional integration is possible only because silicon, besides discussed earlier advantageous electrical and cost/manufacturing related properties, also features outstanding mechanical properties This combination is unique to silicon and cannot be reproduced using any other material.

In broad terms, MEMS/NEMS devices fall into two categories of microsensors and microactuators. In the former case mechanical motion of the parts of the MEMS device caused, for instance, by acceleration (accelerometers), or pressure (pressure sensors) is converted into electrical signal. In the latter case, MEMS device is converting electrical signal into mechanical motion by engaging micro-motors, micro-gears (Fig. 3.25), and other mechanical parts comprising its structure.

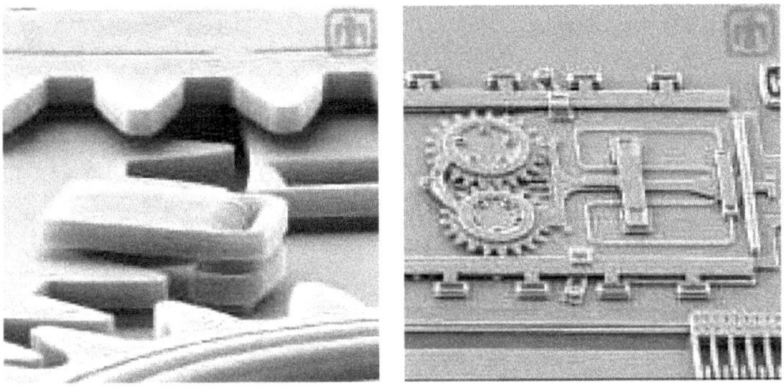

Fig. 3.25. Examples of the MEMS devices processed into silicon (Sandia National Laboratories).

Silicon is the best suited material for MEMS applications for two reasons. First, moving silicon pieces such as cantilever beams or membranes, when flexed and released, return to their original state without dissipating any energy. Process can be repeated virtually unlimited number of times without measurable material fatigue. Second, silicon is a material allowing integration of micro-electrical (MEMS) and electronic (integrated circuits) functions on the same chip using existing readily available manufacturing methods. As a result, silicon-based MEMS/NEMS technology continues to grow driven by the increasing complexity and functionality of MEMS/NEMS components being part of the complete Systems-on-Chip (SOC).

Critical to the operation of any MEMS device are its moving parts shaped into cantilever beams, membranes, and others. Processing such moving parts into solid silicon substrate requires dedicated operations which are specific to the MEMS fabrication procedures and known as the MEMS release processes, (Erdamar *et al.*, 2008). Figure 3.26 illustrates in the simplified fashion the process employed to form and to release a cantilever beam in the silicon-based MEMS device using silicon dioxide as a sacrificial layer.

The ultra-small MEMS devices integrated on the same chip with controlling them electronic circuitry constitute yet another key technical domain driving progress in semiconductor science and engineering. MEMS devices are used in virtually in any and all technically involved consumer products such as conventional cars (e.g. accelerometers controlling airbag deployment, stability control, tire pressure), autonomous cars (e.g. gyroscopes and inertial navigation systems), airplanes (e.g. autopilot functions),

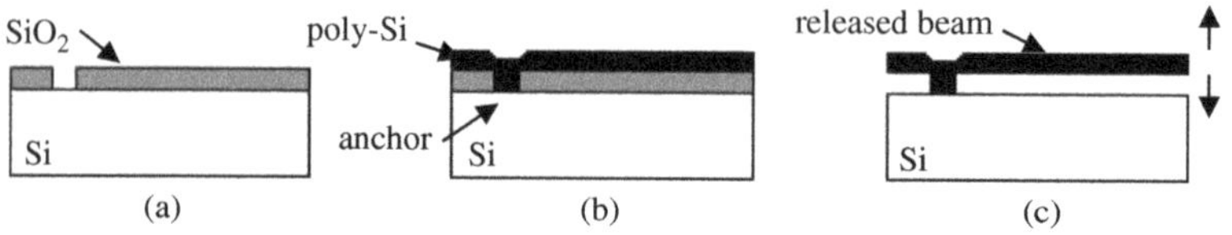

Fig. 3.26. Simplified illustration of the MEMS release process as a result of which moving cantilever beam is formed, (a) silicon wafers covered with patterned layer of SiO_2 acting as a sacrificial oxide, (b) deposition of polycrystalline silicon, and (c) beam release process by sacrificial oxide etching using anhydrous HF:alcoholic solvent vapor mixture.

drones, computer games controllers, smartphones, energy harvesting systems, Digital Light Processing (DLP) systems used in video projection, fitness tracking devices and others. Special role is played by the bio-MEMS devices allowing Lab-on-Chip technology which often takes advantage of unusual behavior of fluids in micro-channels and micro-chambers explored by the science and engineering of microfluidics.

Sensors: The action of *sensing* is understood here as an ability of the physical object to detect in real time changes in its physical/chemical environment and to translate these changes into a measurable signal such as an electric current. As discussed earlier, semiconductor materials and devices are uniquely suitable for sensing applications because, unlike metals and common insulators, several physical characteristics of semiconductors change in response to the changes of the physical or chemical characteristics of the ambient.

In the case of semiconductors, sensing can be accomplished by exploiting changes in semiconductor material properties when exposed to an ambient affecting these properties. As an example, Fig. 3.27 shows the use of the MOSFET with gate oxide fitted with the gas-sensing coating as a chemical sensor. Here, rather than by responding to the changes of the voltage applied to gate contact, the effect of the ambient on the surface potential within channel is used to control output drain current I_D of the transistor, and thus, to respond to the changes in the gaseous ambient composition and density.

From the applications standpoint, there are several distinct classes of semiconductor sensors. In addition to mechanical sensors (see moving beam in Fig. 3.26 which can respond to acceleration) and chemical sensors (Fig. 3.27), there are magnetic sensors responding to magnetic field, radiation sensors, thermal sensors, as well as acoustic sensors just to

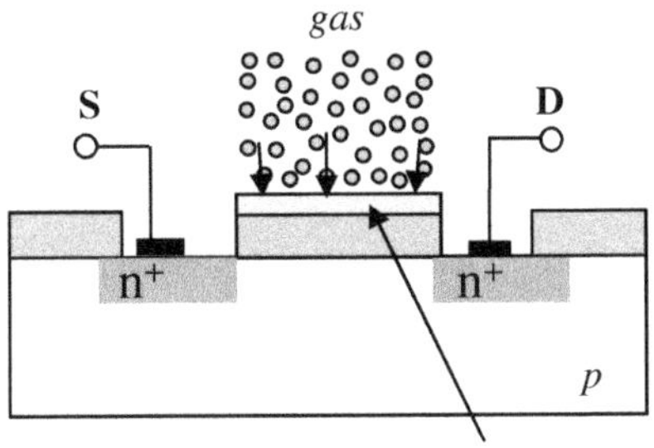

Fig. 3.27. MOSFET in which metal gate is replaced by the gas sensitive coating is acting as a gas sensor.

mention the most important classes of semiconductor sensors. A very special role among semiconductor sensors is played by bio-sensors in which semiconductor-based systems provide unique solutions in medical prevention, diagnostics, and monitoring.

3.9 Wearable and Implantable Semiconductor Devices

Portable electronic equipment such as radios, tape recorders, radio communication devices and others were made possible by replacing vacuum lamps which much smaller and lighter, and using much less power, semiconductor diodes and transistors. Subsequent progress in semiconductor technology made possible portable information processing equipment such as laptops and information transmitting equipment such as cellular phones. Then came time where all of the above, again taking advantage of the progress in semiconductor technology, could be interconnected wirelessly into what has become an internet of things, or IoT.

In the light of the developments stressing portability, mobility, and on-the-go accessibility, permanent or temporary integration of ultra-light and ultra-low power, high-end semiconductor-based electronic and photonic devices and systems with our bodies and clothes we are wearing continues to be aggressively pursued avenue of growth. It goes beyond what we can carry in our pockets (smartphones, notebooks, etc.) and is concerned with instrumentation that is integrated with our clothes or attached to our body for the purposes of interacting with our bodily functions.

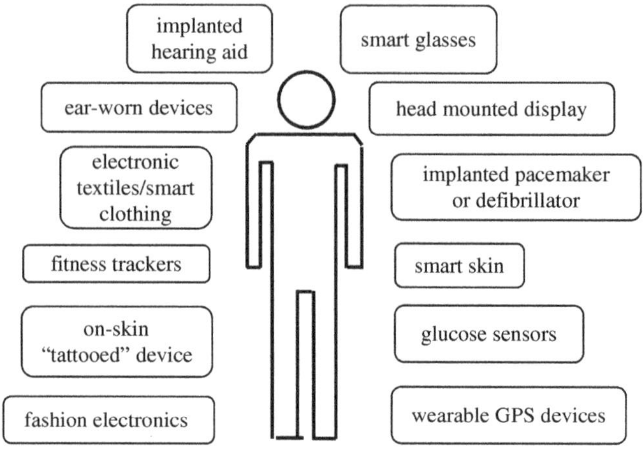

Fig. 3.28. Examples of wearable technology devices.

The comments in this section are meant to recognize wearable semiconductor electronics and photonics as increasingly important part of the broadly understood technical domain of semiconductor engineering. It is in this spirit that Fig. 3.28 gives examples of wearable semiconductor-based electronic instruments selected to point out four classes within wearable technology identified for the purpose of this discussion.

First class is concerned with devices which, in contrast to the devices such as smartphones that we carry around merely out of convenience, are designed to interact with our bodies. Fitness trackers and smart glasses are representatives of this class of devices. The former, in many aspects using MEMS sensors technology, provide detailed information regarding fitness-related activities of the user. Smart glasses in turn are designed to create for the wearer a sense of "augmented reality" by projecting computer screen on the lenses of the glasses in front of wearer eyes. Here, computational capabilities of semiconductor-based electronics are combined with image processing features of semiconductor imaging devices.

The second group is concerned with stretchable and washing resistant semiconductor devices and circuits permanently integrated into fabrics. The very concept of electronic textiles (e-textiles) and smart clothing significantly enhances a range of applications of wearable electronics and photonics. E-textiles are textiles that incorporate adequately shaped and processed computing electronic circuits, LEDs, and sensors woven together with clothing fibers.

Third group includes semi-permanent, or permanent on-skin devices taking advantage of the unique in terms of flexibility characteristics of organic semiconductors which make them suitable for integration with highly flexible and stretchable human skin. As an example, flexible electronic tattoo acting as a sensor, for instance, can be printed directly onto human skin. Such electronic smart skin devices can be designed to monitor health of stroke patients, heartbeat of sick babies, or monitor sweat biochemistry of people involved in sports.

Finally, somewhat special in the context of wearable electronics is a class of implantable devices installed in our bodies for the purpose of monitoring and/or stimulating certain vital bodily functions. Among the range of implantable electronic devices artificial cardiac pacemakers and hearing aids in the form of inner ear implants are obvious examples.

The function of wearable semiconductor-based systems is best illustrated by considering equipment and instruments embedded in or attached to the uniforms of soldiers who must tackle variety of challenges encountered on the battlefield. These include threat detection, GPS systems, health monitoring, night vision, identification tags, communication systems including displays, as well as solar cells to mention just a few.

Besides humans, it is a common practice to implant microchips for animals' identification and tracking purposes. Important point is that whatever the operational principle of an implanted device is, none would be functional without built-in, ultra-low power, miniature semiconductor based electronic circuitry.

Review Questions

3.1. What are two most important electronic semiconductor devices?
3.2. What are two elements needed to construct active semiconductor device?
3.3. Point out key differences between electric diodes and photonic diodes. Give examples of both.
3.4. Explain working of semiconductor solar cell and show its output characteristic.
3.5. Explain working of the BJT.
3.6. Explain working of the MOSFET. Is it a unipolar or bipolar device?
3.7. Define threshold voltage of the MOSFET.

3.8. What is gained by integrating *n*-type and *p*-type MOSFETs into complementary pair (CMOS)? What is gained by structuring MOSFETs this way?
3.9. How does FinFET varies from the conventional planar MOSFET?
3.10. List special characteristics of thin-film transistors. How do they differ from conventional FETs?
3.11. What are two major classes of integrated circuits distinguished based on the function they perform?
3.12. Show schematic diagram illustrating multilevel metallization applied in integrated circuits.
3.13. Your phone is equipped with image sensing and image displaying semiconductor devices. Explain working of both.
3.14. What are the two major categories of MEMS devices and how corresponding MEMS devices are used?
3.15. Give example of three wearable/implantable instruments operation of which is based on semiconductor devices.

Chapter 4

Fabrication of Semiconductor Devices

4.1 Introduction

Semiconductor process technology, understood as a technology developed and used for the purpose of manufacturing semiconductor devices, involves complex tools, methods, and procedures devised specifically for semiconductor device fabrication (Jeager, 2001; Nishi and Doering, 2008). Semiconductor manufacturing is unique as some devices, for instance, cutting edge logic integrated circuits, are among very few mass-produced objects manufactured with atomic scale precision. As a reminder justifying this assertion, let's recall as an example 10 nm long gates in cutting edge logic CMOS circuits expending over some thirty silicon atoms, and even thinner quantum well layers in ultra-high frequency transistors. In addition, there are nanoscale semiconductor material systems such as one atom thick graphene, or less than ten atoms in diameter nanodots, which are readily available commercially.

This chapter addresses the semiconductor device processing related issues which are general in nature and not specific to the types of materials involved (Geng, 2018). Instead, discussion in this chapter is concerned with the substrate geometry dependent process implementation, process media, process environment, and configuration of tools used in semiconductor processing all affecting manufacturing yield and device reliability (Kuo and Kim, 1999).

Keeping in mind objectives and scope of this *Guide*, discussion in this chapter is concerned with principles of semiconductor manufacturing processes focusing on the wafer substrates with only occasional comments in reference to other types of substrates identified in Section 4.2. Besides being very common in semiconductor device manufacturing, rigid wafer substrates are a good platform upon which discussion of semiconductor device process technology can be based.

Furthermore, processes applied in semiconductor device manufacturing are considered in this chapter from the point of view of energy used to stimulate semiconductor fabrication including thermal, plasma, as well as electron and ion beam processes.

In terms of process environment in semiconductor device manufacturing, issues related to contamination control are considered including nature of contaminants encountered in semiconductor fabrication facilities and the ways to keep them under control by employing cleanroom infrastructure and process integration.

4.2 Substrates

The way various processing step involved in the manufacture of semiconductor devices are implemented depends on the size, shape, degree of flexibility, and chemical makeup of materials used to construct such devices. This overview of substrate related issues in semiconductor device technology is an extension of the discussion in Section 2.8 based on which three different ways of substrate dependent process implementation are distinguished.

4.2.1 *Wafer Substrates*

Most commonly, a piece of semiconductor material used to fabricate functional devices is in a form of a thin wafer, typically thinner than 1 mm, featuring single-crystal structure. More detailed discussion of various types of semiconductor wafers and related fabrication procedures was presented earlier in this volume. Here, selected practical aspects related to the use of wafers in the manufacture of semiconductor devices are summarized.

To begin with, a distinction needs to be made between small and large wafers as well as square/rectangular and circular wafers all with the choice of desired surface orientation. In terms of size, commercial circular wafer substrates can be as small as 20 mm in diameter in the case of some II–VI compound semiconductor materials, and as large as 450 mm in diameter in the case of silicon wafers. Clearly, while the nature of processes wafers are subjected to during device manufacturing remains the same regardless of the size of the wafer, the way various operations are implemented, and the way wafers are handled depend on the size and the shape of the wafers. For instance, square, and as thin as some 50 μm Si wafers used in the manufacture of solar cells, where cost of material rather than mechanical stability of the substrate is an issue, are subject to different handling procedures than close to 1,000 μm thick, 450 mm in diameter circular Si wafers where mechanical rigidity of the substrate is a prime concern.

In the following discussion, considerations of the wafer-based manufacturing procedures will be limited to the mainstream manufacturing processes which are using rigid circular wafers. It is postulated that the processes carried out on those type of wafers in an adequate way represent different methods and methodologies employed in semiconductor device fabrication.

The substrate wafers in semiconductor technology are getting larger in order to increase number of devices built into the wafer and in consequence to decrease the cost of electronic functions performed by those devices. Economics of semiconductor processing are such that as large as possible number of identical devices, or chips, is fitted onto the wafer. As Fig. 4.1(a) shows, initial bare wafer after being subject to the hundreds of precisely executed operations is transformed into the wafer containing an array of functional devices, confined within a single chip (Fig. 4.1(b)). Following completion of the fabrication sequence wafers are separated into individual chips which are then packaged and used as a part of the electronic circuitry according to their specifications.

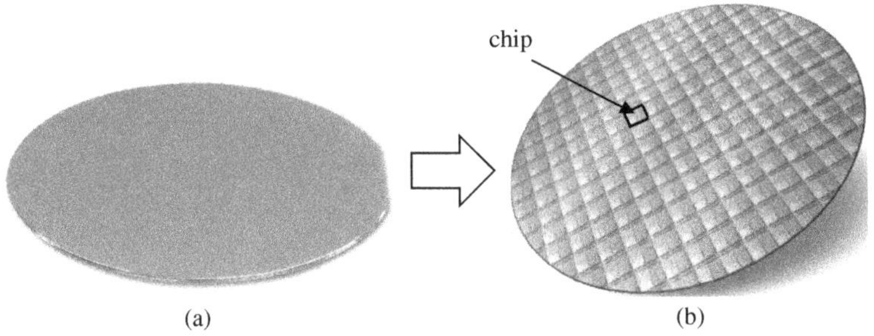

Fig. 4.1. From the bare semiconductor wafer to the processed wafer with hundreds of self-contained individual devices/chips.

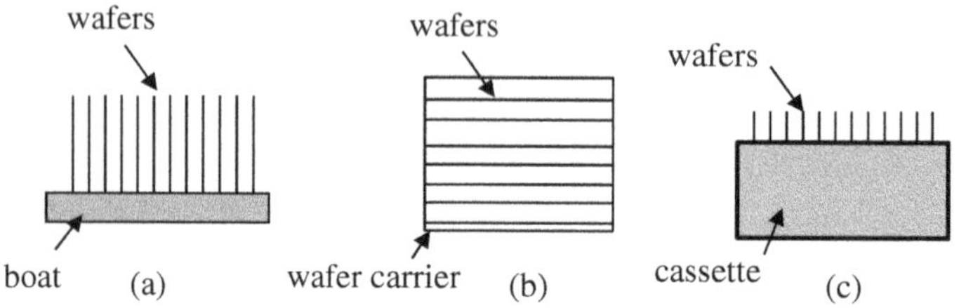

Fig. 4.2. Schematic representation of the different ways batch processes in semiconductor manufacturing are implemented.

In terms of process implementation, the choice is between single-wafer processes where one wafer at the time is exposed to the process ambient, and batch processes (Fig. 4.2) where the number of wafers is loaded into the process tool in batches and then processed simultaneously. As the discussion in Chapter 5 will reveal, there are some manufacturing operations which cannot be executed on the wafers in batches, but otherwise fabrication steps involved in semiconductor device manufacturing can be implemented either in batch, or in the single-wafer process mode.

The choice between arrangements of wafers processed in batches (Fig. 4.2) depends on the nature of any given process. For instance, arrangement where the wafers are loaded vertically into the boat (Fig. 4.2(a)) is often employed in high-temperature processes which include deposition/growth of various thin films. On the other hand, arrangement shown in Fig. 4.2(b) is often encountered in low-pressure processes while the batch process involving chemically resistant cassette depicted in Fig. 4.2(c) is at the core of wet processing where wafers are immersed in liquid chemicals and water.

Advantage of the batch processes is in the higher manufacturing throughput expressed in terms of the number of wafers processed per hour by any given tool. On the other hand, any process failure in this case affects the entire batch and multiplies resulting losses. In the case of single wafer processes the impact of process malfunction is not only limited to a single wafer, but also because each wafer can be monitored individually, it can be detected in real-time. Also, as the wafers are getting larger, they are getting thicker in order to assure required mechanical stability needed to prevent any deformation, and thus, heavier (450 mm wafer weights about 200 grams, while its 300 mm and 200 mm counterparts weight about 125 and about 50 grams, respectively). Increased weight of the load makes high-precision robotic handling of the wafers in batches more demanding. Taking all the above into consideration, it is not surprising that the gradually increasing number of processing steps in advanced semiconductor manufacturing is implemented in the single-wafer mode.

It needs to be noted that the above discussion of rigid, wafer-shaped substrates not exceeding some 450 mm in diameter (or equivalent area in square/rectangular configuration) is not limited to the single-crystal semiconductor wafers only. As discussion earlier in this book points out, there are other substrates used to manufacture semiconductor devices, e.g. sapphire wafers, processing of which is subject to the similar rules and constraints as in the case of semiconductor wafers.

4.2.2 *Large Area Substrates*

A distinct challenge in semiconductor device fabrication is concerned with extra-large glass panels used to manufacture large displays. Connection to semiconductor technology comes from the fact that the key element involved in the display manufacturing is a sequence concerned with a fabrication on the TFT (see Section 3.5.6). In Active-Matrix displays TFTs are used to power up image creating components of the display either in liquid crystal (liquid crystal display (LCD)), or Light Emitting Diodes (LED) display technology. In principle, in the processing of large substrates the same in nature as in the case of wafer-based manufacturing processes are employed, however, their implementation is modified to accommodate significantly larger substrates.

To better understand the challenges of TFT manufacturing technology on extra-large glass substrates, let's recall 450 mm in diameter Si wafer surface area of which is 0.161 m^2 and the weight is about 0.2 kg. By comparison, 10th generation "mother" glass substrate is 2.85 m × 3.05 m and features surface area of 8.693 m^2 which makes it over 50× larger than the largest Si wafer. The weight depends on the thickness and type of glass, and for the largest glass planes it is in the range of dozens of kilograms. It means that that semiconductor device manufacturing involving extra-large glass substrates requires much bigger tools and adjusted accordingly manufacturing infrastructure as compared to the conventional wafer-based manufacturing. What remains the same is the driving force behind the push toward larger substrates which in both cases is aimed at lowering the price of the final product. Furthermore, the general approach to the production scheme in these two cases is the same. In both cases the entire substrates are subject to processing and only after completion of the entire sequence wafers are separated into individual chips, while large glass substrates are cut into smaller panels which after additional processing are converted into the large TV screens.

4.2.3 *Flexible Substrates*

The functionality of semiconductor devices and systems is expanded when in addition to their fabrication on the rigid substrates, they are also available in flexible variations. As the discussion in Chapter 3 indicated, semiconductor devices find applications as flexible electronic circuits, flexible displays, solar cells, lighting panels, as well as elements of wearable electronics and photonics. Obviously, implementation of flexible semiconductor electronic and photonic devices requires not only flexibility of the materials used to fabricate devices, but also adequately flexible substrates upon which such devices are constructed.

As indicated earlier, there are various types of flexible substrates used in semiconductor technology. The point is that flexible substrates call for the significant modifications of the manufacturing technology as compared to the processes performed on the rigid substrates. It means that the same operations such as deposition or pattern definition used in conjunction with rigid substrates must be adapted to the needs of flexible substrates such as, for instance, plastic foil.

Depending on application, plastic substrates are processed either as stationary separate sheets, or as ribbons of foil moving between unwinding roll and the rewinding roll in the process known as roll-to-roll (R2R) process. As Fig. 4.3 illustrates, foil moving between the rolls is exposed to the operations needed to form device features.

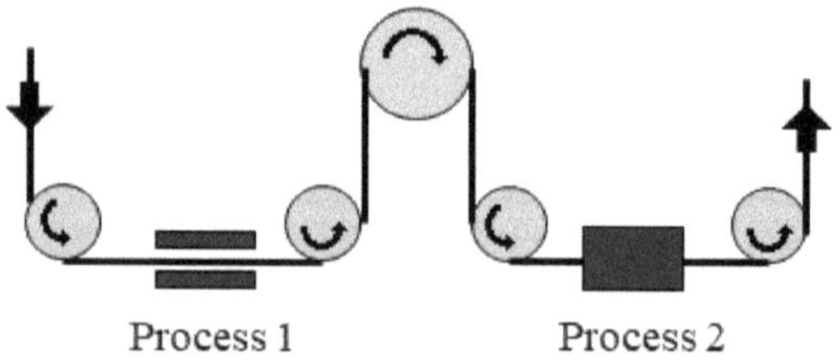

Fig. 4.3. Schematic illustration of the roll-to-roll (R2R) process employed in manufacturing of semiconductor devices on flexible substrates such as plastic ribbons.

In the typical commercial R2R processes several rolls are involved allowing for multiple operations performed sequentially on the moving foil. In this way multi-layer material systems comprising semiconductor devices can be formed. In its essence, the R2R process is similar to the operation of the newspaper printing press.

4.3 Liquid-Phase (Wet) Processes

In spite of being relatively expensive and demanding from the point of view of required infrastructure and safety, as well as in spite of environmental concerns, processes involving liquid chemicals and water (commonly referred to as *wet processes*) have not been entirely eliminated from semiconductor device manufacturing environment. In semiconductor processing wet chemistries and water are used for the purpose of cleaning of the processed materials, as well as in materials etching operations.

Considering variety of materials and differently shaped substrates used, there are various ways of wet processes implementation in semiconductor manufacturing. What also needs to be pointed out is that some of the substrates and process implementation techniques are not compatible with conventional liquid-phase processes. Moreover, the surface tension and possibility of bubbles formation may obstruct the penetration of the fine, high aspect ratio surface features by liquid agents.

In the case of operations compatible with wet processes, common approach involves immersion of processed wafers loaded in the cassette (Fig. 4.2(c)) in liquid agents with spraying used as an alternative solution. Regardless of how any given wet process is implemented, liquid chemicals of desired composition diluted in water are the key element in any wet-phase process.

4.3.1 *Water*

Water is a medium of special importance not only in wet processing of semiconductor wafers, but also in the other parts of semiconductor manufacturing endeavor (for instance, in cooling applications) where it is by far the largest consumable. In fact, water consumption per wafer exceeds by the factor of at least three consumptions of energy, elemental gases, and specialty chemicals combined.

In the course of device manufacturing water serves three main purposes. First, water is an agent which is used to establish desired composition of the chemical solutions used to process wafers. Second, water is used to stop chemical reactions by overflowing reactive chemistries to which semiconductor wafers are exposed. Finally, water is an agent used to remove products of chemical reaction from the surface through the process of rinsing.

In semiconductor device processing only the highest purity water from which all ionized organic and inorganic minerals and salts are removed is used. Such water is referred to as deionized (DI) or demineralized (DM) water. The measure of water purity is its electrical resistivity. Water, not exposed to any ambient, is considered to be absolutely pure at 25 °C when its resistivity reaches 18.2 MΩ-cm at which point it is controlled solely by H^+ and OH^- ions and as such is significantly purer than the distilled water featuring resistivity in the range of 0.5 MΩ-cm and drinking water with the resistivity of less than 0.01 MΩ-cm. To put it simply, DI water contains nothing, but H_2O. Even slight departure from the condition of perfect deionization causes decrease of its resistivity. Water used in high-end semiconductor manufacturing must feature resistivity of 18 MΩ-cm to assure adequate performance of various surface treatments requiring interactions with liquids. To accomplish this level of water purity, constant monitoring of the resistivity of water allowing real-time determination of its quality is a standard operational procedure in semiconductor manufacturing.

Process of water deionization is implemented using ion exchange and reverse osmosis installations. Enhancement of water ability to control undesired organic contamination, including bacteria colonies which may develop into particles adhering to the processed surfaces, is accomplished by dissolution in DI water of the controlled amounts of ozone acting as a strong oxidizing agent and producing water known as ozonated water which is broadly used in semiconductor industry.

4.3.2 *Specialty Chemicals*

Large quantities of liquid chemicals are consumed in semiconductor processes involving material removal by means of etching in the liquid-phase or in the surface cleaning operations using liquid chemicals and water. The choices of liquid chemistries used are dependent on the chemical composition of the processed materials and may vary drastically between various solids processed in the course semiconductor device fabrication.

Several types of liquid chemicals broadly used in semiconductor technology include acids such as hydrofluoric acid, HF, sulfuric acid, H_2SO4, and hydrochloric acid, HCl, as well as alkalis including ammonia hydroxide (NH_4OH) potassium hydroxide, KOH, and sodium hydroxide, NaOH. The list of other reactants used in semiconductor processes is relatively long and includes, for instance, hydrogen peroxide, H_2O_2, ammonia fluoride, NH_4F, and silicon chloride, $SiCl_4$. In addition, organic solvents including isopropyl alcohol, C_3H_8O, commonly referred to as IPA and methyl alcohol (CH_3OH, methanol) are important elements of wet processing of semiconductor wafers. In some specialized operations involved in photolithography (see Section 5.5), liquid developers, adhesion promoters, and resist strippers are used.

Just like in the case of water, an issue of decidedly key importance in wet process technology in semiconductor manufacturing is concerned with the purity of chemicals used. Only the highest purity chemicals are suitable for high-end semiconductor manufacturing as any contamination of liquid chemicals with particles and trace metallic impurities may have a catastrophic effect on the performance of the manufacturing process, and eventually on the performance of fabricated devices. The level of contamination is represented by the number of particles of a given size in the volume (mL) of water and liquid chemicals. In the purest semiconductor grade chemicals, the level of contamination with trace metals such as iron, aluminum, or copper is expressed in parts-per-billion (ppb) or parts-per-million (ppm).

In large semiconductor manufacturing facilities to reduce chance of contamination of liquid chemicals, to minimize processing cost, and to improve safety associated with handling of the large volumes of highly aggressive chemicals, a procedure of point-of-use chemical generation is implemented. By following this procedure aqueous chemicals needed are generated by *in situ* (point-of-use) mixing of ultra-pure DI water with gaseous chemicals such as NH_3 or anhydrous (water-free vapors) acids including HF and HCl. Mentioned above processing of ozonated water is yet another example of point-of-use generation of liquid chemical agents.

Concluding brief overview of process chemicals used in semiconductor device technology, it needs to be emphasized that many of them are highly toxic and corrosive liquids and as such must be handled and disposed of with utmost care strictly following established procedures.

4.3.3 *Wafer Drying*

Wafer drying operation is an inherent part of any process carried out in liquid-phase and ending with a water rinse. Water left on the surface to evaporate attracts particles which leave traces on the surface known as water marks. Thus, the process of removing water from the surface, or drying, is a step which determines condition of the surface of the rinsed wafer following any wet treatment.

The very crude wafer drying technique is based on blow drying in which stream of fresh air, or better a stream of dry nitrogen, is directed on the wafer surface. This method is unsuitable for semiconductor device commercial manufacturing processes, but in some situation can be used in research and development. Commonly used in less demanding device fabrication procedures is a method of spin drying in which water is removed from the surface by centrifugal forces during fast spinning of the wafer in clean air.

A drying method of choice commercial processes involving nanometer scale device features is using isopropyl alcohol (IPA) ambient and exploits Marangoni effect. Known as IPA drying, or Marangoni drying the method illustrated in Fig. 4.4 uses the difference between the surface tension of the solid in contact with deionized water and IPA vapor to produce a gradient that is pushing water from the IPA-exposed part of the surface into the volume of water. As Fig. 4.4 shows, surface drying is accomplished by pulling a semiconductor wafer out of the water into the volume of IPA vapor and nitrogen which results in the fast and effective drying of its surface.

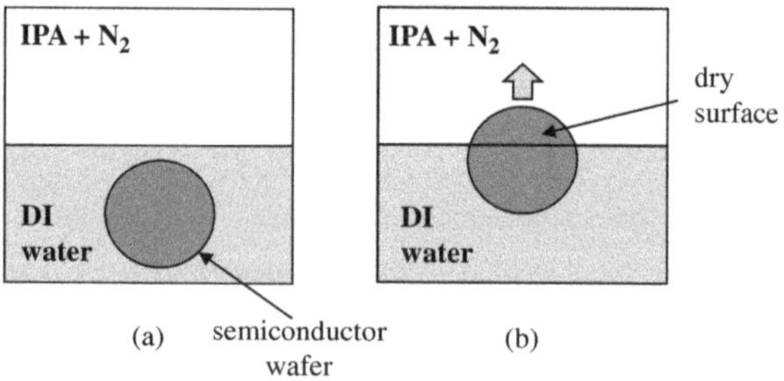

Fig. 4.4. IPA drying of semiconductor wafer based on the Marangoni effect.

4.4 Gas-Phase (Dry) Processes

Processes which are carried out entirely in the gas phase and are commonly referred to as dry processes represent an alternative to the liquid-phase processes employed in semiconductor manufacturing are. Both dry and wet processes are similar in terms of the ability of each to support chemical reactions designed to accomplish specific process goal. For instance, wet processes using hydrofluoric acid HF:water solution can be replaced with anhydrous HF:water vapor solution in certain etching operations (Torek *et al.*, 1985).

In contrast to wet processes which rely entirely on the chemical reactions with the processed material, dry processes allow also physical interactions between species in the gas phase and processed material. Such gaseous species can be electrically charged and can carry significant amount of kinetic energy both of which can be used to enforce directionality (anisotropy) of gaseous ambient interactions with processed material. By relying solely on the chemical reactions, wet processes are by definition highly isotropic. This is one difference in basic characteristics of wet and dry processes which makes the latter more broadly used in semiconductor manufacturing. Others are related to the easier than in the liquids control of contaminations in gaseous ambient (particularly at the reduced pressure), as well as to the challenges related to the cost and infrastructure needed to handle large amounts of water and liquid chemicals in semiconductor manufacturing facilities.

In the following overview of dry processes, discussion is focused on the common gases used in semiconductor processing, as well as on the basics of the gas-phase processes carried out at the reduced pressure and in vacuum.

4.4.1 *Gases*

Essentially any semiconductor manufacturing process requires gases which either provide controlled process environment, or are chemically, or physically in the form of the ions, engaged in the process. Similarly to liquids, also in the case of gases used in semiconductor device processing a chief concern is their purity (Shon, 2015). Only gases featuring highest purity expressed in percent, e.g. > 99.9995% can be used. In some applications gases as pure as 99.99999% are required.

In the following discussion, gases used in semiconductor technology are considered in terms of inert gases which under normal conditions do not engage in any chemical reactions, and process gases, also referred to as specialty gases, which interact chemically with other gases or with solids which are exposed to them.

Inert gases: Inert gases are meant to perform in the gas-phase functions equivalent to the role played by DI water in liquid-phase operations, and thus, need to be chemically inert with respect to the processed materials. While not perfectly chemically inert, nitrogen, N_2, is inert enough to effectively fulfill its role of an ambient gas, carrier gas, and purging gas. Due to its low cost and lack of safety related concerns, nitrogen is universally used across all semiconductor manufacturing processes requiring inert atmosphere. Argon, Ar, is another gas which as a noble gas could play a role of an inert gas in semiconductor manufacturing. However, argon is too expensive to be used in mass applications in the way nitrogen is being used. Instead, argon is commonly used as a chemically inert, but physically active Ar^+ ion in the range of process applications involving discharges in gases discussed later in this volume.

From the practical applications point of view an important characteristic of nitrogen is its relatively high boiling point which allows conversion of gaseous nitrogen into liquid nitrogen (LN) at the industrial scale. Under normal atmospheric pressure, nitrogen can exist as a liquid up to the temperature of 77.2 K (−196 °C). Liquid nitrogen is stored in thermally insulated containers called Dewars from which it is either allowed to evaporate and used as a clean N_2 in semiconductor processes, or remains in the liquid phase and is employed as a cooling agent, for instance in the vacuum systems used in semiconductor research and industry.

Process gases: Process gases also referred to as specialty gases, are the foundation of gas phase, or dry processes in semiconductor manufacturing. The selection of process gas is driven by the specificity of the operation to be performed (e.g. deposition or etching), and chemical composition of materials being processed. In the light of divers needs of various processes, and the broad range of materials used to construct semiconductor devices, the list of process gases used in semiconductor fabrication is long and includes chemically complex gaseous compounds.

Considering multiplicity and complexity of chemical interactions involved, even superficial overview of an extensive array of specialty gases employed in semiconductor engineering is beyond the scope of this discussion. Besides, some aspects of the use of compound specialty gases is considered in Chapter 5 in the context of specific processes involved in semiconductor device manufacturing. Here, only selected key characteristics of hydrogen and oxygen, which are two elemental process gases used across the range of semiconductor manufacturing applications, are briefly summarized.

Consisting of the proton and just one electron, hydrogen is the simplest, lightest, and the most abundant element in the universe, although, rarely appearing in nature as a pure H_2. As highly flammable and potentially explosive gas, hydrogen needs to be handled with utmost attention particularly in terms of its interactions with oxygen, and thus, with air. Depending on the process and processed materials, hydrogen can serve the purpose of an inert gas or a process gas. As such, hydrogen is probably the most versatile gas used in semiconductor device fabrication. As an example, hydrogen is used as a reducing agent during thin-film deposition processes involving, for instance, silicon. It is also used as a carrier gas and diluent for the gallium, arsine, and phosphorus precursors in the manufacture of multilayer III–V devices. In addition, a mixture of 5–15% hydrogen in nitrogen forms so called forming gas used in various process sequences typically in order to stabilize processed semiconductor surfaces.

As a strong oxidizing agent, oxygen is used to form a layer of native oxides on the surfaces of semiconductors such as silicon which lend themselves to oxidation. Oxidizing strength of oxygen is also used to volatilize by oxidation films of organic compounds and to remove residual organic contaminants from the processed surfaces. Furthermore, when exposed to short wavelength radiation, oxygen molecule O_2 breaks apart to produce two oxygen atoms (2O) which then combine with oxygen molecules to produce ozone molecule O_3. High oxidizing strength of ozone is exploited in some processes involved in semiconductor device fabrication including its incorporation into the DI water to form what is known as ozonated water.

Gas flow measurements: Gas flow measurements are an integral part of the gas supply systems in semiconductor manufacturing technology where it is essential to exercise full control over the volume of gas and the rate at which each gas is delivered. The key role in this regard is played by Mass Flow Controllers (MFC) where the flow of gas is measured in standard cubic centimeter per minute (SCCM). The MFCs are constructed and calibrated for the use with specific gas. In the processes less demanding in terms of degree of control, and those using large volumes of gas, the gas flowmeters which measure the flow in liters per minute (l/min) are used.

Reiterating points made in the discussion of liquid process chemicals, it needs to be stressed that also in the case of dry processes extreme caution in the handling of gases used must be exercised. Most of the gases used in semiconductor technology are toxic and corrosive highly volatile substances. As such, they must be handled and disposed of with utmost care strictly following established procedures. With the exception of inert gases discussed earlier and oxygen, essentially in all operations involving process gases, tools using them must be fitted with adequate exhaust systems. Furthermore, facility where they are used must be equipped with efficient gas scrubbing installation designed to decompose process gases before they can be released into the atmosphere.

4.4.2 *Vacuum*

The use of the term "vacuum" in the context of any manufacturing technology is not exactly correct because it refers to the space entirely empty of any matter which as such, would have limited applications in semiconductor manufacturing. Therefore, in practical semiconductor terminology, term "vacuum" is used in reference to the space in which pressure of gases is orders of magnitude lower than atmospheric pressure which as one atmosphere (1 atm) is used in reference to the average atmospheric pressure at the sea level. In the International System of Units (SI), a unit of pressure is pascal (Pa) defined as a force of one Newton per square meter. In the everyday semiconductor process related terminology, however, a unit of pressure more commonly used is Torr. The relationship between one atm, Pa, and Torr is as follows: 1 atm = 1.01325×10^5 Pa = 760 Torr.

Torr is a unit of pressure used consistently in this book where the term high vacuum (HV) is used in reference to the pressure range from 10^{-6} Torr to 10^{-8} Torr, while term ultra-high vacuum (UHV) refers to the pressure below 10^{-9} Torr. Terms such as rough vacuum, low vacuum, and process vacuum are used depending on the context in reference to the pressure range from about 100 Torr (sub-atmospheric pressure) to about 10^{-5} Torr (Jousten, 2016).

Vacuum infrastructure provides an environment especially conducive with the needs of semiconductor manufacturing where purity of the ambient and precise control over various process parameters are of importance. In addition, vacuum is needed to initiate a discharge in gases leading to the generation of plasma which creates unique, and indispensable in semiconductor manufacturing, process environment. Considering all this, it is evident that the vacuum equipment is an integral part of semiconductor manufacturing infrastructure, as well as of any semiconductor related research and development endeavor.

The key elements of any vacuum system are vacuum pumps needed to evacuate air from the process chambers as well as vacuum gauges used to measure pressure in the vacuum equipment. Also, residual gas analyzers (RGA) used to identify residual gaseous species present in the vacuum chambers are an integral part of the typical vacuum system used in semiconductor processing.

Figure 4.5 shows schematic diagram listing types of pumps and gauges used in a typical vacuum system. Among types of pumps shown, distinction is being made between low-vacuum roughing pumps, which

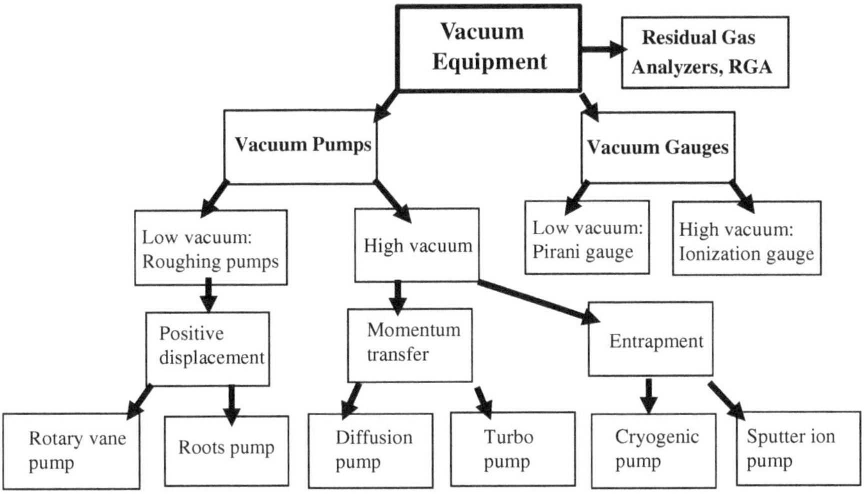

Fig. 4.5. Schematic representation of the components of the vacuum equipment used in semiconductor process tools operating at the reduced pressure and types of vacuum pumps and gauges used.

are used to reduce pressure from atmospheric to low-vacuum level, and high-vacuum pumps which are used to further evacuate gases from the chamber until desired high-vacuum level is reached and then maintain it for as long as needed. Common roughing pumps are positive displacement pumps in which displacement of air from the chamber to the exhaust is accomplished through the rotary motion of their parts. Rotary vane pump and roots blower are examples of this type of pumps.

To pump system down to high-vacuum level requires designated high-vacuum pumps (Fig. 4.5) including momentum transfer pumps and entrapment pumps. Among the former, diffusion pumps are using momentum carrying high speed stream of oil vapor to enforce motion of gas molecules toward the exhaust. Due to the oil vapor back-streaming and contaminating vacuum system, diffusion pumps are rarely used in commercial semiconductor manufacturing equipment. Instead, oil vapor-free, high pumping speed turbomolecular pumps (turbo pumps in short) in which high-speed turbines evacuate gases from the process chamber and direct them toward the exhaust are used. Due to their versatility and performance, turbomolecular pumps are at the core of the typical high-vacuum apparatus used in semiconductor device technology. The other class of high-vacuum pumps is based on the entrapment principle. The most common representative of this class is a cryogenic pump, or cryopump, in which removal of gas is accomplished by adsorption of the gas species on the cold surfaces inside the pump while in the case of ion pump gaseous species are ionized and then trapped at the cathode. Selection of the high-vacuum pump depends on the gases to be evacuated from the system, target pressure of the process, as well as pumping speed required.

Depending on the pressure, two types of gauges are commonly used in vacuum equipment to measure pressure (Fig. 4.5). Low-vacuum Pirani gauges are used in the pressure range from atmospheric to 10^{-4} Torr, while high-vacuum ionization gauges measure pressure in the 10^{-4} Torr to 10^{-8} Torr range.

Due to the lack of efficient pumps working in the pressure range from atmospheric to high-vacuum, typical vacuum semiconductor process tools require two-stage pumping. First, roughing pump, bypassing high-vacuum pump, reduces pressure in the process chamber down to the pressure allowing operation of the high-vacuum pump. Then, high-vacuum pump takes over pumping function while roughing pump gets reconnected to the high-vacuum pump to support its operation. As a side note, exposure of the working high-vacuum pump to the air/gas at the atmospheric pressure, or slightly below, will stall the pump, render it dysfunctional, and will require its costly re-conditioning.

In order to prevent implosion, parts of the high-vacuum system need to be constructed using mechanically sturdy material such as stainless steel, then welded into the larger elements. High-vacuum tools are commonly equipped with electrically controlled pneumatic valves. If due to compromised integrity of the system high vacuum cannot be maintained, the leak detectors are used to localize a weak spot. It is a standard procedure to use liquid nitrogen to cool down certain parts of the vacuum system, most notably some high-vacuum pumps such as diffusion pumps and cryopumps.

4.5 Processes in Semiconductor Device Fabrication

It is intuitively obvious that semiconductor material will sustain no change in its physical or chemical characteristics unless certain amount of energy needed to initiate desired process in its bulk or on its surface is delivered to the wafer. In this section, various ways of delivering energy to the wafers in the course of semiconductor device manufacturing are reviewed. Based on the considerations below, various types of semiconductor processes are identified (Campbell, 2020).

It needs to be noted that different sources of energy can be employed simultaneously when the excessive use of one type of energy, for instance, thermal energy, is not desired. As an example, a given process can be initiated on the surface of semiconductor wafer exposed to short wavelength radiation at a temperature lower than would be required without wafer irradiation.

4.5.1 *Thermal Processes*

Thermal energy is broadly used in semiconductor technology as a stimulant of fabrication processes such as film deposition, dopant introduction, as well as oxidation, and treatments aimed at the modification of the properties of materials comprising a device. By increasing temperature of the wafer, thermal vibrations of the atoms in their lattice sites are enhanced which not only allows structural rearrangements within the solid, but also may promote migration of the alien atoms in the solid. As a result, on many occasions during device manufacturing, temperature of the substrate wafer is increased significantly above the room temperature. The extent of heating depends on the material and the needs of the process, and it may vary drastically. For instance, temperature in excess of 1,200 °C is required to promote oxidation of silicon carbide (SiC), while temperature as low as 200 °C could be the highest to which plastic substrate can be exposed.

In practical applications, temperature is only one element defining the amount of thermal energy delivered to the wafer which is often referred to as thermal budget. The other one is the time of the thermal treatment. If the time of exposure to elevated temperature is very short, in the range of 10–30 seconds, for instance, then even in the case of temperature as high as 1,000 °C the treatment will be considered to be a low-thermal budget process. In contrast, if the wafer would be exposed to 1,000 °C for the time longer than some 20 minutes, then the amount of thermal energy delivered to the wafer will be sufficient to refer to the process as high-thermal budget process. In general, unless specific needs of the process call for long exposure of the wafer to high temperature, low-thermal budget processes are preferred.

Below, various aspects of thermal energy use in semiconductor technology are considered focusing on the most common way of increasing wafer's temperature using radiant heating.

Radiant heating: The concept of radiant heating relies on the use of elements generating large amount of thermal energy which is then transferred to the processed substrate by radiation. The most common heating element is a high resistance wire temperature of which can be raised to the level that depends on the current passing through it. To construct wafers processing thermal reactor, a complete heating element, typically shaped in the form of a cylinder, along with thermal and electrical isolation is packed into the stainless-steel housing. The tube made out temperature resistant, very pure material such as fused quartz, is fitted inside the heating element to act as a process tube. The heating element along with a process tube is mounted in the appropriate frame, and equipped with other instruments necessary to run manufacturing, becomes a resistance heated furnace.

Most commonly, heating elements of the furnace system used to process semiconductor wafers are arranged horizontally to form what is known as a horizontal furnace (Fig. 4.6(a)). To be processed, wafers are inserted into the horizontal process tube in the appropriately shaped boats,

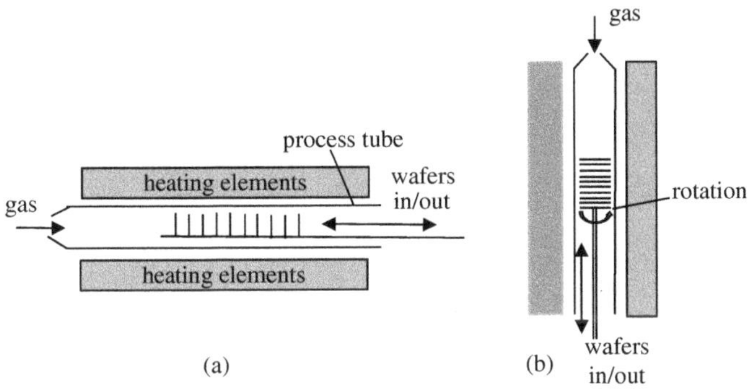

Fig. 4.6. Resistance heated furnace in (a) horizontal and (b) vertical configuration.

made of fused quartz, poly-Si, or SiC, that are attached to, or processed into, the cantilever loading system. This arrangement prevents contact between the boat and the wall of process tube during loading avoiding generation of particulate contamination.

In an alternative configuration process tube is installed vertically to form a vertical furnace (Fig. 4.6(b)). While more demanding in terms of installation, vertical configuration facilitates automation of the wafers loading, allows rotation of the boat with wafers during processing which results in better process uniformity. Vertical configuration also saves space in the fabrication facility due to the smaller than horizontal furnace footprint.

A distinct feature of furnace processing in semiconductor manufacturing is that the time required to complete entire thermal cycle, which includes ramping up wafers temperature to the temperature of the furnace, duration of the process, and finally wafers cooling down cycle, cannot be in practice much shorter than some 15–20 minutes. Therefore, furnaces as described above are not suited for low-thermal budget processes, and thus, differently designed thermal reactors are needed to allow short high-temperature thermal cycles.

The machines used for accelerated thermal treatments are called rapid thermal processors (RTP). Instead of using heavy heating elements used in conventional furnaces, RTPs are commonly using banks of high-power halogen lamps that can be turned on and off instantaneously. When processing one wafer at a time (single wafer process), RTP can increase its temperature at the rate of up to 500 °C/s. In practice, however, such high heating rates are not used to avoid damage of the wafer resulting from

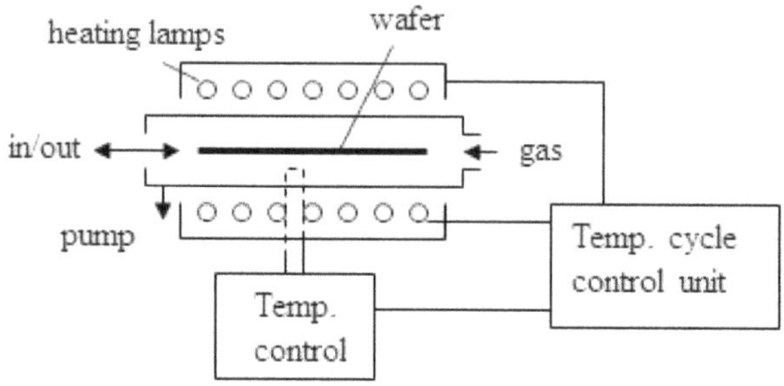

Fig. 4.7. Schematics of the single-wafer Rapid Thermal Processing (RTP) reactor.

thermal stress. Inside RTP, wafer is in the close proximity of the lamps located on one or both sides of the horizontally positioned wafer (Fig. 4.7).

Inductive heating: An alternative to radiant heating is an inductive heating. The concept of inductive heating relies on direct coupling of the energy of high-frequency (microwave in above 50 MHz range or radio frequency, RF, in 100 kHz to 1,000 kHz range) electric field with material to be heated. The microwave heating is widely exploited in consumer microwave ovens operating at significantly lower power levels. Regardless of whether semiconductor wafer is heated directly at microwave frequencies, or indirectly via graphite susceptor in the RF regime, heating in the case of large substrate wafers is not sufficiently uniform and as such may result, for instance, in non-uniform thickness of the film deposited with assistance of inductive heating. Due to this limitation, inductive heating is used primarily in the processing of smaller substrates wafers.

Both radiant heating and inductive heating are uniformly increasing temperature of the entire wafer, and thus, do not allow local heating. Certain processes in semiconductor manufacturing call for the interactions of thermal energy with the substrate to be limited to selected areas on the surface of the substrate. To accomplish local heating, energy arriving at the surface of the heated material must be in the form of the spatially restricted beam. Options in this regard are limited to laser beam heating or electron beam (*e*-beam) heating.

Laser beam heating: In this case, high intensity laser beam is used to deliver energy it carries to the substrate wafer where it is absorbed and converted into heat. Two characteristics of laser heating make it highly suitable for local heating applications. First, laser beam heating is very fast. In the case of conventional material, fraction of the second long laser beam irradiation could be enough to bring its temperature in the irradiated area above its melting point. Second, due to the spatial coherence and spectral purity, the laser beam allows heating that is localized both horizontally and vertically. This means that only selected regions near the surface of the irradiated material are heated while temperature of its remaining part can only be altered due to its thermal conductivity.

The depth of absorption of energy carried by the laser beam depends on the wavelength of laser light and is defined for each material by its absorption coefficient at any given wavelength. The absorption coefficient increases with increased laser photon energy (shorter wavelength). Larger the absorption coefficient, closer to the surface of irradiated material energy of laser light is absorbed. It should be noted that absorption coefficient is different in amorphous and crystalline semiconductors and is also dependent on the dopants concentration and temperature of the irradiated material.

The lasers suitable for semiconductor heating applications include Nd:YAG lasers as well as argon, Ar, lasers. If very shallow heating is desired, a very short-wavelength excimer lasers (e.g. krypton fluoride laser, KrF, with wavelength of 248 nm) featuring high beam uniformity and output power stability can be used. For deep local heating of solids, high-power long-wavelength carbon dioxide, CO_2, lasers (10.6 μm) are particularly suitable.

Electron beam heating: The stream of electrons focused into a fine beam can also be used to heat solids locally. Accelerated electrons carry significant amount of kinetic energy which upon electrons impinging on the solid surface is converted into substantial amount of heat that is dissipated in the near-surface region of the solid. The beam heating is controlled horizontally by the beam size and vertically by the energy of electrons acquired by acceleration. With high acceleration energy and high beam density local melting of bombarded solid can be accomplished. In Section 4.5.3, some basic features of *e*-beam interactions with solids which are exploited in semiconductor technology are considered further.

Thermal energy used in the manufacture of semiconductor devices is the easiest to implement process stimulant and as such is employed across a broad range of semiconductor fabrication operations. However, it needs to be used cautiously and with full understanding of its potential effect. This is because there are several semiconductor materials both crystalline and non-crystalline which cannot be exposed to temperatures above some 600 °C. For instance, compound semiconductors comprised of elements featuring different vaporization characteristics (different vapor pressure) may be subject to decomposition as a result of elevated temperature exposure. Organic semiconductors at the elevated temperature might be outright destroyed just like most of the plastic substrates and some glass substrates.

Furthermore, elevated temperature may bring about redistribution of dopants in semiconductor substrate altering device geometry as well as stimulate undesired surface reactions. It needs to be emphasized that the effect of most of these undesired effects can be minimized through the reduction of the thermal budget of the process by shortening time of exposure to elevated temperature (Rapid Thermal Processing).

As a result of the listed above constraints, the use of sources of energy other than thermal is a necessary practice in semiconductor manufacturing. Energy other than thermal is needed in those applications where heated material features low thermal resistance, as well as in the cases where by supplementing thermal energy with energy from the other sources the desired processing effect can be accomplished at the lower temperature.

4.5.2 *Plasma Processes*

Plasma, often considered as a fourth state of matter, is the most common phase of matter in the Universe both by mass and by volume. As such, it is a key building block of the Universe including gaseous atmosphere surrounding Earth but does not occur naturally on Earth where it needs to be artificially generated. It displays unique properties which when exploited in the controlled fashion, make plasma a powerful tool in numerous technological applications, semiconductor processing including (Shul and Pearton, 2000).

Generation of plasma: Plasma consists of partially ionized by the electric field gas containing equal positive and negative charges, as well as gas molecules which remain unionized, and thus, electrically neutral. Positive charge is contributed by ions of atoms losing an electron in the ionization process such as, for instance, argon ion Ar^+ ($Ar - e^- \rightarrow Ar^+$). Negative charge is contributed by the free electrons present in the plasma and playing crucial role in the ionization process. In addition, negative charges can

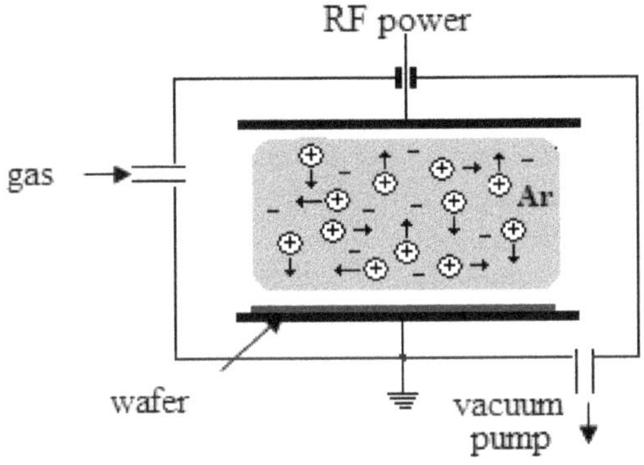

Fig. 4.8. Plasma generated in the parallel plate reactor.

be contributed by the species originating from gas molecules such as for example Cl_2 which upon decomposition in an electric field feature unpaired electron, and thus, carry a negative charge ($Cl_2 \rightarrow 2\ Cl^-$). Species featuring unpaired electrons are called free radicals and are chemically more reactive than in the corresponding non-radical form.

Plasma is a result of electrical discharge in the gas at the reduced pressure typically in the range from 0.01 Torr to 10 Torr. As an example, Fig. 4.8 illustrates in the simplified fashion process of plasma generation in low-pressure argon in the conventional parallel-plate reactor.

When high electric field is applied to the gas confined between electrodes (anode and cathode), free electrons will be released by the field emission either from gas molecules, or from electrodes. These electrons are accelerated by the electric field and eventually gain sufficient energy to ionize gas molecules by inelastic collisions. The process creates ions and additional free electrons, and the chain reaction spreads ionization process across the entire gas volume, although, under typical conditions, only a small fraction of the gas atoms is ionized during the discharge. The degree of ionization is expressed in terms of plasma density representing percent of ionized atoms and is typically less than 1% with the remaining atoms maintaining electrical neutrality. In the self-sustained processes occurring in plasma, atoms are continuously ionized through collisions with electrons then rapidly de-excited releasing energy in the form of the bright glow (glow discharge). The wavelength of emitted light is determined by the composition of gas in which plasma is generated.

When positioned on the electrode inside plasma chamber (Fig. 4.8), semiconductor wafer is subjected to various interactions with the species within the plasma. Type of these interactions depends on the gas, its pressure, (Beck, 1980), configuration of the plasma reactor, and, most importantly, on the way electric field is applied to the volume of gas to ignite plasma. Preferred in most applications, including semiconductor processing, is the use of alternating current voltage (AC voltage, AC plasma) applied between the electrodes. The frequency of alternate signal can vary from low kHz range to GHz range covering the entire spectrum of radio frequencies (RF plasma in Fig. 4.8 generated at 13.56 MHz is an industrial standard), and reaching up to microwave range (microwave plasma at 2.45 GHz is a standard frequency in this case).

High-density plasma: In practical applications of plasma in semiconductor fabrication processes, low density plasma (below 1%) does not allow efficient plasma-stimulated processes. Therefore, high density plasma (HDP) reactors are common in semiconductor manufacturing. The increase of plasma density is accomplished by combining the gas-ionizing effect of electric field with the effect of magnetic field which confines free electrons within the plasma, and thus, leads to the increased ionization efficiency which in turn results in the increased density of plasma. As an example, Fig. 4.9(a) shows schematic representation of the Inductively-Coupled Plasma (ICP) reactor. Due to the inductive coupling, in contrast to capacitive coupling in the case of parallel-plate reactor in Fig. 4.8, the ICP reactor allows positioning of the plasma generator outside of the process chamber. In this way electric and magnetic fields are efficiently combined toward increased plasma density. Alternatives configurations generating high-density plasma involve helicon plasma and Electron Cyclotron Plasma (ECR plasma) reactors.

In some cases, direct exposure of processed material to the high-energy radiation associated with plasma, as well as to energetic ions

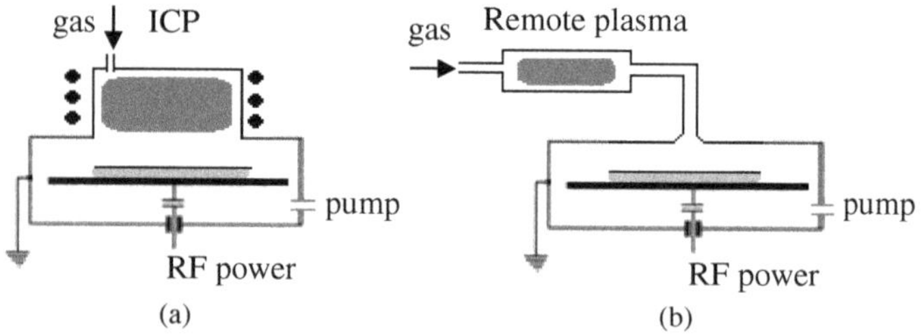

Fig. 4.9. (a) Inductively coupled plasma (ICP) and (b) remote plasma reactors.

originating from plasma may be harmful to the processed material itself, or to the features formed on its surface. To cope with this issue remote plasma, also known as downstream plasma reactors are employed in semiconductor manufacturing processes. As Fig. 4.9(b) shows, in remote plasma reactor configuration substrate wafers are exposed to energetically relaxed plasma afterglow instead of being directly exposed to plasma and its products as in the case of the parallel-plate reactor (Fig. 4.8).

The very merit of plasma processes in semiconductor technology can be summarized in the following three points. First, plasma generates species which at the same time are electrically charged and may feature high chemical reactivity. As a result, chemical reactions involved plasma generated ions can be directed using electric field. Second, plasma allows more reactive process chemistries and at the lower temperature than thermal stimulation only would allow. As a result, plasma enhanced (PE) processes are commonly used to lower temperature of the processes involving, for instance, thin-film deposition. Finally, plasma is a source of ions which, as discussion later in this section reveals, by themselves and not as the ingredients of plasma, are used in selected semiconductor fabrication processes such as ion implantation (see Chapter 5 for more detailed considerations).

In conclusion, in semiconductor processes plasma is a manufacturing "tool" which generates unique environment for the promotion of chemical reaction and physical interaction with material undergoing processing. All this makes plasma a process medium which plays a pivotal role in semiconductor fabrication technology.

4.5.3 *Electron and Ion Beam Processes*

The process in which kinetic energy carried by electrons and ions in motion is released upon their impinging on the surface of the solid is broadly exploited in semiconductor device technology. The energy is imparted to electrically charged electrons or ions by acceleration in the electric field. The amount of energy carried defines the effect electrons or ions have on the bombarded solid surface and near-surface region. Whatever the effect is, it can be localized over the limited area of the solid surface by focusing either electrons or ions in motion into a beam.

Considering that (i) the mass of a typical ion is some five orders of magnitude larger than the mass of an electron, (ii) the size (atomic radius) of the ion is some thirteen orders of magnitude larger, and (iii) electrons penetrating the solid are annihilated through recombination while ions may remain in its structure as the permanent modifiers of its chemical composition, the nature of ions and electrons interactions with bombarded solids is obviously very different. Therefore, the way they are used in semiconductor processing is also different.

Electron beam processing: The goal of using electron beam (*e*-beam) in semiconductor technology is twofold and is distinguished by the beam current which depends on the number and velocity of electrons forming the beam. First is the case of high current *e*-beam which, as discussed earlier in this section, involves local heating of the solid by bombarding electrons. Second is concerned with electrons forming low-current beam which rather than heating of the bombarded material is causing changes of its chemical makeup. For instance, properly formulated polymers exposed to electron beam will experience either breaking of the existing intermolecular bonds, or formation of the new bonds by the process known as crosslinking. This effect is a foundation of the patterning technology employed in semiconductor manufacturing known as *e*-beam lithography (see Chapter 5 for additional discussion).

To avoid collisions of electrons accelerated toward the substrate with residual gas species in the process chamber, all *e*-beam processes must be carried out under the high-vacuum conditions. Also, generation of free electrons which form a beam can only be effective in the high-vacuum environment regardless of whether source of electrons is a hot cathode or cold cathode.

Available electron optics systems allow focusing of the *e*-beam down to less than 1 nm in diameter. However, due to their very low mass, electrons penetrating the solid are subject to severe scattering due to the collisions with atoms comprising that solid (Fig. 4.10(a)). The first among scattering effects involves forward scattered electrons which are displaced from their original direction by less than 90 degrees. The second concerns backscattering which can displace electrons in motion by as much as 180 degrees. In this last case, electrons cannot only be returned to surface, but also, provided they preserve enough energy, can leave the solid. Moreover, some inelastic collisions of incident electrons with atoms in the solid can result in the emission of secondary electrons from this solid.

The combined effect of the electron scattering related phenomena is such that the area over which electrons energy is absorbed in the solid is

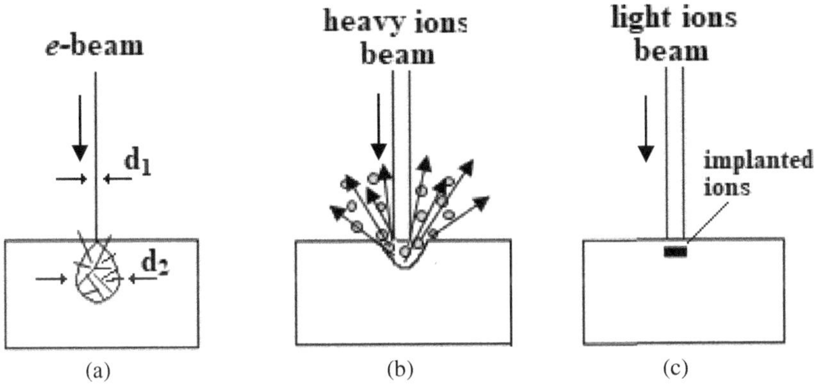

Fig. 4.10. Interactions of electron and ion beams with the solids: (a) *e*-beam and electron scattering, (b) heavy ion beam and sputtering, (c) light ion beam and ion implantation.

larger than the cross-section of the incident beam itself ($d_2 > d_1$ in Fig. 4.10(a)). The phenomena preventing exact reproduction of the *e*-beam geometry in the solids are referred to as proximity effects.

Ion beam processing: The concept of using ion beam as a process tool in semiconductor device technology is similar to that of using *e*-beam. However, as indicated earlier in this section, ions or in other words atoms or molecules that have lost one or more electrons, are much heavier than electrons. As a result, the effect of accelerated electrons and ions impinging on the surface of the solid is very different. Therefore, unlike light electrons which upon entering the solid are subject to scattering and eventual annihilation through recombination, the ions are penetrating the solid colliding with the host atoms. In the case of heavy ions acquiring through acceleration energy in excess of bonding energy of bombarded solid, atoms from the bombarded solid surface are ejected in the process known as sputtering (Fig. 4.10(b)). Lighter ions penetrating solid do not cause ejection of the host atoms and instead are losing energy through collisions and come to the rest at certain distance from the surface and remain there (Fig. 4.10(c)). This behavior is a foundation of an important material processing technique known as ion implantation.

As the discussion in Chapter 5 indicates, both sputtering and ion implantation are the effects exploited in semiconductor device manufacturing.

In addition to ions energy acquired through acceleration, also geometry of the ion beam has an impact of the outcome of the process. The requirements of specific application determine the use of either showered ion beam (SIB), or focused ion beam (FIB). The former is used predominantly for the dopants introduction by implantation and material removal by sputtering, while the latter can be used in the direct-writing of desired patterns on the solid surfaces. It needs to be noted, however, that because of their mass, ions cannot be focused into as fine beam in terms of cross-section as electrons can.

Regardless of the application, ions require electric discharge in gases (plasma) to be generated from which they are extracted and accelerated toward the target material as either showered or focused ion beam.

4.5.4 *Chemical Processes*

Number of processes important in semiconductor manufacturing rely on chemical reactions carried out either in the gas phase, or in the liquid-phase. Besides thermally and plasma stimulated chemical interactions considered earlier, as well as photochemical interactions to be discussed later in this section, there are chemical reactions which are driven by the energy available within the chemical system in the amount sufficient (activation energy) to initiate chemical interactions meant to achieve desired process goal. The goals can be very diverse and may include surface cleaning, material removal, or thin-film deposition.

In general, chemical reactions employed in materials processing involve complex changes in the energy balance of the chemical system depending on whether breaking of the bonds, or formation of the new bonds is the goal. For instance, some chemical reactions, known as exothermic, release energy in the form of heat while others, known as endothermic, take energy from the environment. Still others, can be both exothermic and endothermic.

Additional considerations regarding chemical processes employed in semiconductor manufacturing are interwind in the remaining discussion in this chapter, and then in Chapter 5. More detailed considerations of the complex chemical reactions involved in semiconductor processing are beyond the scope of this introductory discussion. What needs to be recognized here, however, is that chemistry is at the very core of materials processing toward fabrication of functional semiconductor devices.

4.5.5 *Photochemical Processes*

Photochemical processes in semiconductor manufacturing are using short wavelength light/radiation to stimulate desired chemical reactions in the illuminated solid, gas, or in the liquid. The term short wavelength is used here in reference to the light in the UV (ultraviolet) portion of the electromagnetic spectrum featuring wavelength λ shorter than 400 nm and extending all the way to the wavelength as short as 10 nm which is a "soft" X-ray range (Fig. 4.11).

At such short wavelengths the UV light carries significant amount of energy which increases with decreasing wavelength from 3.1 eV at 400 nm to as high as 12.4 eV at 10 nm. In contrast to discussed earlier longer wavelength light (infrared) featuring heat generating absorption mechanism in solids, the prime role of short wavelength light is to induce photochemical reactions in the illuminated medium by changing its chemical properties. An example of such process is a photolysis which refers to the decomposition of chemical compound, usually in the gas-phase, under the influence of light. To achieve such decomposition, the energy of photons must be higher than the energy of chemical bonds in the irradiated compound. In these terms, photolysis is an analogy to the process of thermal decomposition (pyrolysis).

Depending on the UV range, different sources of UV light are used. In the near UV range (Fig. 4.11) halogen lamps are employed as UV light sources. Dedicated UV gas-discharge lamps containing different gases, as well as excimer lasers are needed to produce UV light in the deep UV range. Elaborate very high-density plasma sources are required to generate extreme UV radiation most commonly featuring 13.5 nm wavelength.

The most important use of UV light in semiconductor manufacturing is in pattern defining photolithography processes considered in more

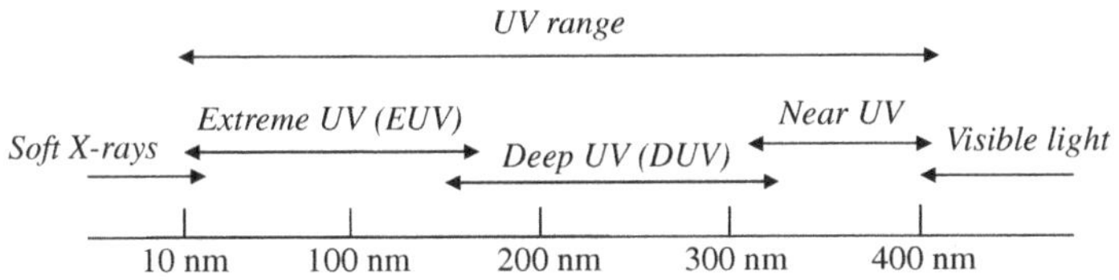

Fig. 4.11. The UV portion of the electromagnetic spectrum and its parts identified using terminology adopted in semiconductor nomenclature.

details in Chapter 5. Besides photolithography, photo-stimulated processes offer alternative solutions in applications in which elevated process temperature is prohibited, as well in those requiring reduced pressure plasma tools which bring about increased complexity to the process infrastructure. In general, use of photochemical processes is driven by the simplicity of instrumentation used and very low thermal budget. For instance, UV stimulated formation of ozone, O_3, which is a strong oxidizing agent used in semiconductor processes for oxidation of hydrocarbons, is accomplished by exposing air, or pure oxygen to UV light containing 185 nm and 254 nm wavelengths in its spectrum (Ruzyllo, 2025). The UV light breaks oxygen molecule into two atomic oxygens which then react with oxygen molecule to form ozone ($O_2 + UV \rightarrow 2O$ then $O_2 + O \rightarrow O_3$).

4.5.6 *Chemical-Mechanical Processes*

In yet another embodiment of chemistry use in semiconductor manufacturing, strength of the chemical reactions is combined with mechanical forces in the Chemical-Mechanical Polishing/Planarization processes broadly known as CMP. As name indicates, CMP processes are used to remove top-surface layers of the processed material through polishing, or to eliminate unevenness of the surface features through the planarization process. The process uses chemically active slurry containing nanosized abrasive particles in conjunction with polishing pad and mechanical forces to enforce chemical-mechanical interactions resulting in the gradual removal of the top layers of the processed material (Fig. 4.12).

As Fig. 4.12 illustrates, the top surface of the turned upside down wafer is pressed against polishing pad and the slurry, while wafer holder and polishing pad are counter rotated. Composition of the polishing slurry and pad material are selected depending on the material being polished.

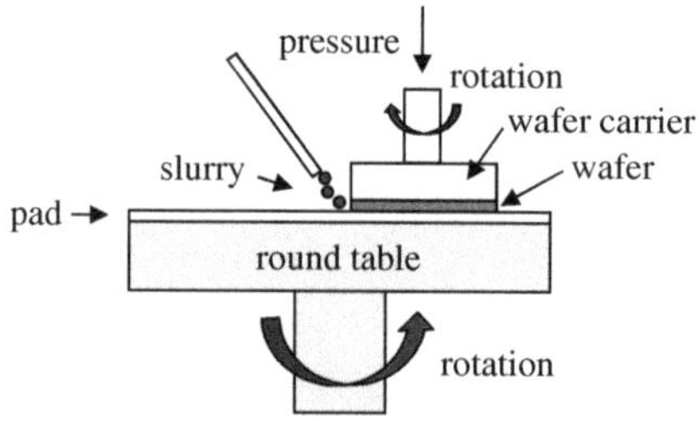

Fig. 4.12. Simplified representation of the CMP process.

The desired rate of material removal is additionally controlled by the applied mechanical pressure and rotation speed.

The CMP processes are common in semiconductor processing in a variety of applications ranging from the thinning of wafers and material removal to the elimination of the surface roughness of the processed materials (see additional discussion concerning CMP processes in Chapter 5).

4.6 Contamination Control

In semiconductor device fabrication contamination of the production environment, including tools and media used, has a profound adverse effect on the manufacturing yield and product performance. There are several reasons for which in semiconductor technology utmost attention needs to be paid to the control and prevention of contamination with the following two being the most evident. First, with features of some semiconductor material systems being processed with atomic-scale precision, any interference of even tiniest contaminants can have a catastrophic effect on the outcome of the process. Second, very high sensitivity of the properties of materials used to construct semiconductor devices to outside influences such as contamination, in combination with elevated temperature employed at various stages of device processing, as well as frequent exposures to varied ambients from which alien species (contaminants) are difficult to eliminate, creates circumstances making semiconductor processes particularly vulnerable to contamination.

In this section, contaminants of particular concern in semiconductor engineering are identified and then measures which are taken to minimize possibility of contamination interference with semiconductor manufacturing processes are briefly considered.

4.6.1 *Contaminants*

Before listing specific contaminants, bear in mind that different types of contaminants interact differently with different materials under different process conditions. For instance, surface contamination with metallic species would be a major issue in the case of high temperature process involving silicon in the manufacture of advanced logic integrated circuits, but would be of lesser concern in the case of surface contamination of the glass used as a substrate in organic solar cells processing. Therefore, consideration of the role of specific type of contamination needs to be considered in the context of specific materials and processes.

Particles: The contaminants which have an unquestionable adverse effect across all semiconductor process technologies are particles and particulates (groups of particles) adsorbed at the surface of the substrates being processed. Particles contamination in semiconductor processing may originate from the multiplicity of sources including ambient air, liquid chemicals, water (where the bacteria colonies end up acting as particles), and gases, as well as process tools and wafer handling operations (Hattori, 1998). In addition, people running fabrication facility are adding particles to the surrounding air, mostly skin flakes, in large quantities. Size of particles vary, depending on the source, from as large as 10 μm to as small as 100 nm, and smaller. Problem is that particles as small as that are difficult to detect and visualize, but their effect on some processes and manufacturing yield can still be catastrophic. In fact, size of the particles is a factor differentiating between the extent of their impact in various processes. For instance, if allowed on the silicon surface, 0.5 μm in diameter particles would ruin any advanced, nanometer-scale integrated circuit manufacturing process. The same size and chemical makeup particles could have less adverse effect in the manufacture of the large area silicon solar cell panels, for instance.

Particles on the wafer surface have an adverse effect on the device performance at every stage of device fabrication. For instance, particles allowed on the bare silicon surface will alter properties of the near-surface region during subsequent elevated temperature processes such as thermal oxidation or epitaxial deposition. In turn, particles contaminating patterned wafer surface later in device fabrication sequence (Fig. 4.13), are disrupting any subsequent process and eventually affecting device manufacturing yield.

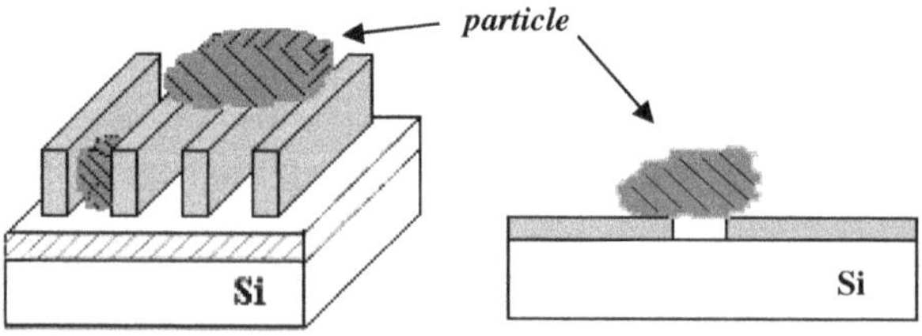

Fig. 4.13. Particle on the patterned silicon surface interferes with subsequent process regardless of the device fabrication sequence.

Overall, presence of particles in any semiconductor process environment is highly undesired and as discussion of cleanroom technology later in this section will show, no efforts are spared to minimize their harmful effect. Because total prevention of particle contamination of processed wafers is for all practical purposes not possible, elaborate methods aimed at the removal of particles during wafer cleaning operations needed to be available (see discussion in Section 5.4.1).

Organic contaminants: Organics are compounds of carbon with other elements most often with hydrogen forming hydrocarbons C_xH_y. Under normal conditions hydrocarbons are present in essentially any environment including ambient air, liquids, and gases. In some situations, accumulation of organic contaminants adsorbed on the processed surface may lead to process malfunction. For instance, organics allowed on the surface will interfere with subsequent deposition processes, and in the case of metal deposition will have a harmful effect on the properties of electrical contact to the surface.

Organics cannot be eliminated from ambient air, including ultra-clean cleanroom air (see later discussion), and their adsorption at the processed surfaces cannot be entirely prevented. Therefore, focus is on the removal of organics which is based on oxidation and requires relatively simple procedures (see discussion of surface cleaning in Chapter 5).

Metallic contaminants: The sources of metallic contamination in semiconductor processing include primarily liquid process chemicals and water. They may also contaminate processed surfaces because of the physical contact with metal parts during wafer handling operations. Furthermore, undetected corrosion of the metal parts in water or gas supply lines may result in contamination of the surface of the processed wafers with metallic species. Among metals, the most aggressive contaminants, each in its own way, are metals such as iron (Fe), copper (Cu), and aluminum (Al) (Fig. 4.14(a)) Listed metallic contaminants on silicon surface subjected to high-temperature treatment such as thermal oxidation will have a different deleterious effect depending on how any given metal interacts with silicon and its native oxide (silicon dioxide, SiO_2) during the high-temperature oxidation (Fig. 4.14(b)).

Temperature activates metallic contaminants on the wafer surface and in some cases promotes their penetration into the substrate wafer (Cu in Fig. 4.14(b)) where they create electrically active defects altering in a

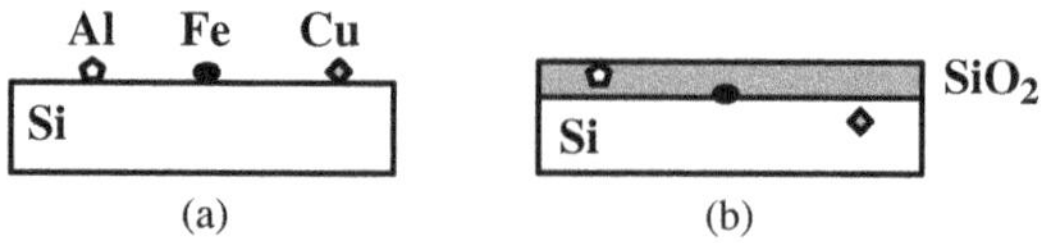

Fig. 4.14. (a) Common in semiconductor process environment metallic contaminants on Si surface and (b) segregation coefficient driven distribution of the same metallic species in Si-SiO_2 after thermal oxidation.

major way transport of charge carriers in affected region. Iron (Fe) in turn remains on silicon surface during thermal oxidation increasing density of traps at the Si-SiO_2 interface while aluminum (Al) atoms will remain close to the surface of the growing oxide contributing negative charge to the oxide.

Their deleterious effect of metallic contaminants extends across the broad range of devices particularly those which require elevated temperature exposure during manufacturing. For that reason, and despite the low level of contamination of process chemicals and water with alien metal atoms in state-of-the-art manufacturing environment, in some wafer cleaning sequences applied prior to high temperature processes in particular, steps aimed specifically at the metallic contaminants removal are required.

Moisture: Strictly speaking, moisture is not a "contaminant" in the way mentioned above particles or metallic contaminants are. However, moisture adsorbed on any solid surface has a destabilizing effect by promoting chemical reactions which would not be initiated in the absence of moisture. The matter is particularly challenging when it comes to interactions between moisture and organic contaminants of the wafer surface. Moisture promoted chemical reactivity is accompanied by electrical interactions as dissociation of water leaves H^+ and OH^- ions on the surface ($H_2O \rightarrow H^+ + OH^-$). For that reason, control of moisture in semiconductor process environment, as well as the time of wafers exposure to such environment need to be carefully executed.

4.6.2 *Clean Environment*

Considering sensitivity of semiconductor materials and devices to contamination, an ultra-clean process environment needs to be assured to accomplish economically viable manufacturing yield. The concept of "clean environment" encompasses both media (gases, chemicals, water), as well as facilities in which semiconductor manufacturing is taking place.

Process media and tools: Materials used in the manufacture of semiconductor devices must be the highest grade available in terms of purity. The issue of purity of process chemicals, water, and gases was addressed earlier in this chapter. The same requirements apply to solid precursors used in deposition processes as well as to substrates used in device manufacture.

In addition to the process media, tools and processes on their own often generate contaminants which may end up being adsorbed at the wafer surface. For instance, particulate contamination could be a product of some etch processes. Besides, essentially all tools used in semiconductor manufacturing, robotic wafer handlers, for instance, involve moving parts. Unless designed specifically to cope with friction-resulting contamination, moving parts within the tool will generate large amounts of particles of various nature. Furthermore, materials used to construct process chambers designed to carry out processes employing aggressive chemical reactions must maintain their chemical integrity at elevated temperatures. As a matter of fact, elevated temperature itself may cause outgassing of contaminants from the process chamber even without any chemical reactions involved. The same applies to materials used to fabricate cassettes, boats, containers, and other parts used to handle wafers during processing.

An issue of importance in semiconductor industry is concerned with the storage boxes and shipping containers used to store wafers between various processing steps in device manufacturing sequence, as well as those used to ship wafers from, for instance, wafers manufacturer site to the device manufacturing facility. Commonly used containers tend to outgas organic compounds which end up being adsorbed on the surfaces of the shipped wafers. These effects must be accounted for by dedicated cleaning operations performed on the wafers before starting manufacturing process.

Cleanroom: A well-known concept of a "cleanroom" refers to the enclosed environment in which semiconductor device manufacturing processes, as well as other contamination-sensitive industrial manufacturing processes, must be carried out to assure satisfactory outcome of the fabrication sequence. Industrial cleanrooms provide restricted access spaces in which at least parts of the semiconductor manufacturing sequences are carried out.

The main purpose of the cleanroom is to create an environment which, to the degree dictated by the process needs, is particulate contamination-free. Also, cleanrooms are designed to assure strict control over temperature, moisture content in the ambient air (typically set at 45%), and static electricity generated by the motion of the large volumes of filtered air recirculated through the manufacturing facility. As a rule, air pressure inside the cleanroom is maintained slightly above atmospheric pressure such that no unconditioned air from outside the cleanroom can penetrate inside the cleanroom.

Among several measures used to define particle contamination level in the cleanroom air the one most common in everyday usage divides cleanrooms into classes from class 1 to class 100,000 depending on the number particles of a given size per cubic foot of air. Accordingly, representing the cleanest air class 1 allows only one 0.5 μm particle per cubic foot of air, and no particles larger than 0.5 μm. The "dirtiest" class 100,000 cleanroom allows 1,000,000 particles sized at 0.5 μm and up to 700 particles featuring 5.0 μm per cubic foot of air. For the manufacture of integrated circuits featuring low-nanometer geometries, cleanrooms class 1, or better (class sub-1) are indispensable. At the same time class 1,000, or even class 10,000, could be enough to carry out parts of the manufacturing processes involving devices featuring relaxed geometries.

Depending on the cleanroom class, either HEPA (High Efficiency Particulate Air) or ULPA (Ultra Low Penetration Air) filters are used. In conjunction with a laminar flow of the air recirculated through the filters (as opposed to turbulent flow which in terms of particles handling is more disruptive), HEPA filters are suitable for the class 100 and higher cleanrooms, while ULPA filters are needed for cleanroom class 10 and below.

Due to the disruptive role of people inside the cleanroom, participation of human personnel in cleanroom-based operations is restricted as much as possible. Semiconductor processing creates an unusual situation in which human operators and engineers are isolated from the environment by wearing sophisticated protective clothing (gowns) not only to protect themselves from the often-hazardous process environment, but mainly to protect process environment against the people-generated contamination.

Depending on the type of semiconductor devices manufactured (e.g. high-density integrated circuits, solar cells, or large area displays) configuration of the cleanroom facilities may vary significantly. Examples of two cleanroom layouts representing different needs of semiconductor manufacturing processes are schematically illustrated in Fig. 4.15.

The first one (Fig. 4.15(a)) represents a layout known as ballroom cleanroom in which any given cleanroom class is maintained in the entire cleanroom and in which process tools are installed, along with most of the process supporting infrastructure, inside the cleanroom. Some tools in the ballroom cleanroom can be isolated into so-called minienvironments featuring better than the rest of the cleanroom class. For instance, in the class 1,000 ballroom cleanroom a minienvironment featuring class 100 can be installed around selected process tools carrying out operations which are particularly sensitive to particulate contamination (Fig. 4.15(a)).

Due to the sheer volume of the air that needs to be processed in the ballroom cleanroom, maintaining the class 1,000 in this type of installation is challenging. Attempts to create even cleaner environment would be difficult to justify in light of the cost of such installation on one hand, and availability of the cleanroom layout solutions better compatible with class 10 or better process needs on the other hand.

One such solution is represented in Fig. 4.15(b) by so called bay/chase cleanroom configuration. Here, the high cleanroom class such as class 1 is maintained only in the limited footprint bays from which wafers are loaded, through the properly configured load locks, into the process tools installed in the service chase. At the chase side wafers are isolated from the ambient air, and thus, class 1000 is enough to support service-related operations that need to be performed on the tools and process infrastructure.

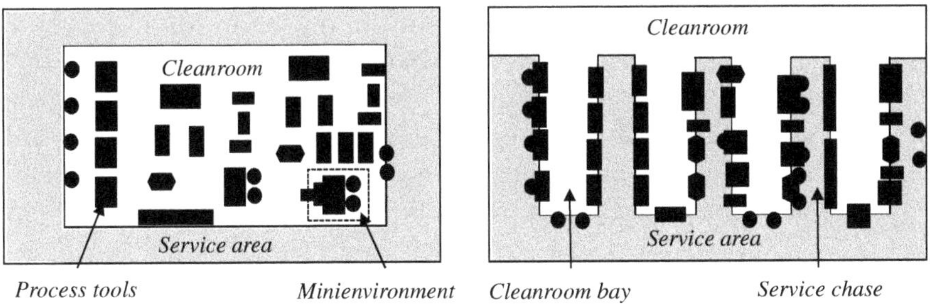

(a) (b)

Fig. 4.15. Common cleanroom layouts (a) ballroom cleanroom, (b) bay/chase cleanroom.

The cleanroom-based major manufacturing facilities at the class 1 or below level are complex and expensive in terms of capital investments. As such, they are justified in high volume manufacturing of only some types of semiconductor devices such as low-nanometer scale digital integrated circuits. Less demanding in terms of air control needs cleanroom facilities, such as class 1,000 and above, are commonly used both in research and development, as well as in less demanding in terms of air control industrial manufacturing.

4.7 Process Integration

There are various ways semiconductor manufacturing processes are being implemented in terms of how individual tools performing designated operations are organized into the manufacturing lines. For instance, large scale fabrication of solar cells involves conveyor belt-like installations moving processed substrates between process stations. Also, processes carried out on the rigid-wafer substrates, while involving same types of operation such as thin-film deposition, for instance, are executed in a different way than roll-to-roll processes carried out on the plastic ribbons in motion (Fig. 4.3).

As an example, we shall briefly consider the processes performed on rigid substrate wafers in the high-end integrated circuit manufacturing facilities. Specifically, the concept of process integration using cluster tools is considered by comparing it with standard approach using standalone process tools.

The way in which processed wafers are transported between standalone manufacturing tools performing designated operations is determined to a significant degree by the size, and thus, weight of the wafers, as well as the class of the cleanroom facility. In the case of smaller wafers, manual transport of the wafers enclosed in the air-tight autopod wafer carriers, often using push carts, serves the purpose. The autopod wafer carriers are SMIF (Standard Mechanical Interface) compatible and are designed to provide minienvironment with controlled airflow and pressure in which transported wafers are isolated from any contamination. In the case of larger wafers and more elaborate manufacturing sequence, essentially the same SMIF pods carrying wafers are transported between process tools by Rail Guided Vehicles (RGV), or Automatic Guided Vehicles (AGV) moving on the cleanroom floor. In the most advanced manufacturing facilities, for instance, class 1 cleanrooms featuring limited-space process bays (Fig. 4.15(b)), an Overhead Hoist Transport (OHT) system moving pods with wafers between process tools is often employed.

The need to use complex wafers transport systems is reduced and handling of the wafers minimized when instead of stand-alone tools each performing single operation, individual process modules are integrated into multi-module cluster tools. In such integrated tools more than one operation can be performed on the wafer without breaking the vacuum or otherwise exposing it to varied ambients and handling/transport related hazards.

In addition to limiting wafer transport and handling, cluster tools require smaller footprint of class 1 process bays in the cleanroom as compared to distributed tools arrangements. Also, vacuum, gas supply, and process control infrastructure serving each module in the distributed system, is centralized into a single system in the case of the cluster.

The concept of process integration is illustrated in Fig. 4.16 using simple sequence involving three processing steps as an example. In Fig. 4.16(a), three separate process tools, each performing dedicated operation, require multiple wafers loading and unloading steps and transfer of the wafers between process modules using one of the wafer transport schemes introduced above.

Figure 4.16(b) shows process tools performing the same operations installed around the wafer handler platform very often using standard MESC (Material and Equipment Standards and Code) ports. Shown in this figure is the simplest four-sided cluster. In real-life industrial

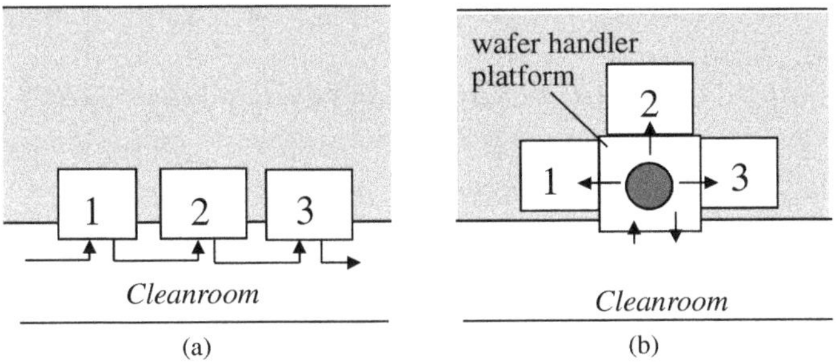

Fig. 4.16. (a) Wafer processing using stand-alone tools installed in the service area, (b) the same processes integrated in the cluster tool.

manufacturing situations six-sided clusters, and even eight-sided clusters are used. In some cases, multi-module clusters are integrated into even larger clusters.

In the light of the listed above benefits of "clusterization", the trend toward increased degree of process integration in semiconductor manufacturing will continue. With time, remaining challenges related to the execution of certain processing steps in the cluster environment will be worked out, at which point integration and automation of the entire fabrication sequence involved in the manufacture of integrated circuit will be a viable option.

Review Questions

4.1. What types of substrates are used in fabrication of semiconductor devices? What understood under the term "wafer substrate"?

4.2. How pure water needs to be, in terms of resistivity, to be used in semiconductor device fabrication?

4.3. Name and explain physical phenomenon upon which commonly employed in semiconductor device fabrication wafer drying method is based.

4.4. List components of the typical vacuum system used in semiconductor manufacturing.

4.5. What methods are used in the manufacturing of semiconductor devices to increase in the controlled fashion temperature of the processed wafers?

4.6. Explain the process of plasma generation and the purpose of plasma use in semiconductor device fabrication.

4.7. For what purposes electron beam can be used in semiconductor device fabrication?

4.8. In semiconductor device fabrication removal of top part of material (etching) can implemented by bombardment with high energy ions. Name and describe the process.

4.9. UV light is broadly used in semiconductor device processing. Describe its main use in this application. What is UV wavelength range used in this case?

4.10. What is a CMP process and for what purpose it is being used I semiconductor device manufacturing?

4.11. List contaminants commonly encountered in semiconductor device fabrication environment which, if allowed on the surface of the wafer, would have adversely impact the process and its results. How would it happen? Describe the effect of various contaminants.

4.12. Consider layouts of cleanrooms where fabrication of semiconductor devices is taking place.

4.13. What is a measure of cleanroom air quality in cleanroom where fabrication of semiconductor devices is taking place. Is clean room class 100 cleaner than cleanroom class 10?

4.14. Give an example of a specific processing sequence integrated into single tool (cluster tool).

4.15. In your mind what are the features of semiconductor device/integrated circuit fabrication distinguishing it from other commercial fabrication procedures? Do you have to wear special clothing in former case. If so, why?

Chapter 5

Fabrication Methods in Semiconductor Device Technology

5.1 Introduction

Following general introduction to semiconductor process technology in Chapter 4, this chapter reviews fabrication methods used to modify semiconductor material in such a way that the final product is a functional semiconductor device. The discussion is based on the fabrication procedures involving silicon wafers which are used here as a representation of the methods employed in mainstream semiconductor device manufacturing technology in ultra-clean environment (Kozicki *et al.*, 1991).

The review of semiconductor fabrication processes in this chapter is concerned with operations which transform bare semiconductor wafer into discrete semiconductor devices or integrated circuits in the top-down manufacturing scheme (Van Zandt, 2014; Vossen and Kern, 2012; Yoo, 2008). A common top-down semiconductor processing sequence of operations performed on the wafer during device manufacturing is identified and then followed in the ensuing discussion. It includes coverage of the key patterning methods, surface preparation techniques, additive and subtractive processes employed in semiconductor manufacturing, as well as methods used to alter semiconductor material's properties by adding alien elements to its structure by so called doping. The chapter is concluded with an overview of procedures involved specifically in the back-end-of-line operations including contact and interconnect processing and the assembly and packaging techniques.

5.2 Pattern Definition Schemes

The key element in the semiconductor engineering process, which aims at the fabrication of the functional semiconductor devices, is the creation of intricate patterns on the surface and near-surface region of the processed substrate. Depending on the type of substrate used (for instance, rigid wafer or flexible foil), and materials comprising a device (for instance, single-crystal silicon or organic semiconductor), appropriate to the needs pattern definition scheme from among those considered below is implemented. In the first three techniques listed, deposition of the film to be patterned and pattern establishing step are applied in sequence. In the other two, using shadow masks or printing, pattern definition and thin film deposition are implemented in the single step. The last one, namely stamping, involves options requiring somewhat different considerations.

In state-of-the-art semiconductor technology pattern defining techniques span across a broad range of physical, chemical, photochemical, and mechanical interactions and as such escape clear-cut grouping into well-defined classes of processes. Therefore, the listing adopted in the following discussion is somewhat arbitrary and serves the purpose of illustration of various trends in patterning technology rather than all-encompassing definite classification of various pattern definition methods used in semiconductor device engineering.

5.2.1 *Top-Down Process*

The essence of the top-down sequence is that the thin film to be patterned is deposited on the entire substrate (blanket deposition) first, and the patterning process follows. In simplified terms a top-down process can be explained as follows.

The first step in the sequence is deposition of the thin film of the material to be patterned (Fig. 5.1(a)) using one among several thin-film deposition techniques discussed later in this chapter. Once deposited, film to be patterned is coated with thin layer of the photosensitive material known as photoresist (Fig. 5.1(b)).

Subsequently, photoresist is selectively illuminated through the mask comprised of the transparent to UV light parts, and non-transparent opaque parts blocking off UV light (Fig. 5.1(c)). The opaque and transparent parts reflect the pattern to be created in the layer of photoresist. Following UV exposure, wafer is immersed in the solution known as a

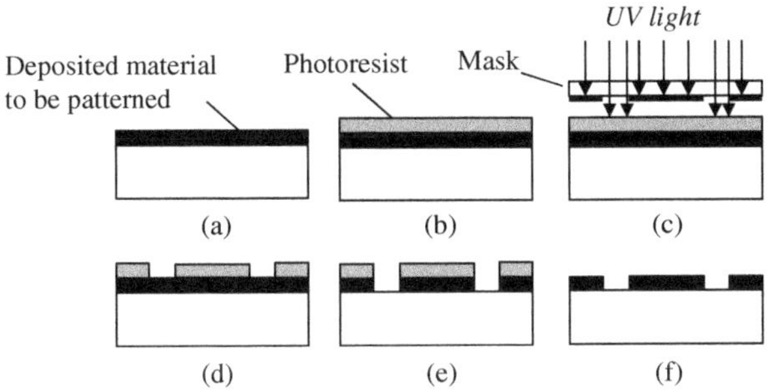

Fig. 5.1. Processing steps creating patterns on the surface of semiconductor wafer in the conventional top-down sequence.

developer which dissolves parts of photoresist which were exposed to UV light, or those not, exposed depending on the type of developer used. At this point desired pattern is transferred to the substrate wafer, but for now it is created in the pattern transfer layer of photoresist, and not yet in the thin-film underneath (Fig. 5.1(d)). To accomplish the latter, wafer is subjected to the etching: process using chemistries which remove material in open spaces while not attacking photoresist (Fig. 5.1(e)). The process is completed with photoresist stripping operation and cleaning (Fig. 5.1(f)). Upon completion of the top-down patterning sequence illustrated in Fig. 5.1, wafer is ready for further processing.

The sequence of steps described above is the most common pattern defining scheme applied in fabrication of semiconductor devices.

5.2.2 *Bottom-Up Process*

There are materials which cannot be patterned using top-down sequence discussed above because of structural fragility or solubility in process chemicals involved in the patterning process rendering them incompatibility with photoresist processing and etching operations. In such cases an alternative scheme involving bottom-up patterning sequence is employed.

The fundamental difference between top-down and bottom-up processes is that in the former case the patterning step is performed on the film deposited on the surface of the wafer while in the latter, patterning step locally changing chemical makeup of the surface is carried out first and then the material is deposited following pattern created on the surface.

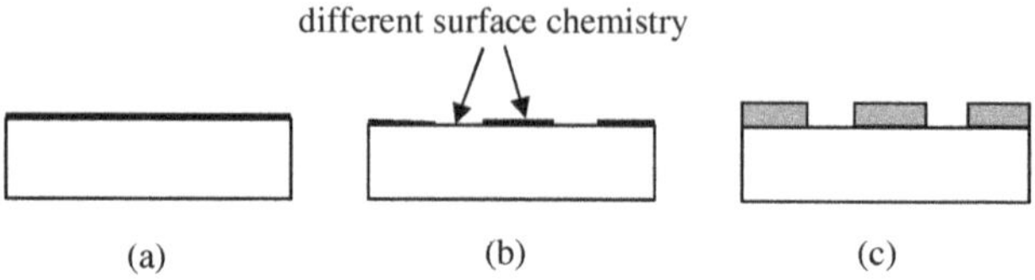

Fig. 5.2. Processing steps creating patterns using bottom-up sequence, (a) surface functionalization, (b) patterning, chemical functional group remain in selected areas, and (c) bottom-up growth of the material following pattern established during step (b).

The bottom-up patterning sequence can be implemented in a variety of ways. As an example of the bottom-up approach, a process in which surface is first covered with a chemical compound serving the purpose of surface functionalization by altering surface energy is shown in Fig. 5.2(a). Then, the surface is processed, using, for instance, localized UV exposure such that chemical functional groups remain on the surface only in places where the material to be patterned is expected to be formed (Fig. 5.2(b)). During the subsequent step, parts of the surface featuring different surface energy will be coated with a thin-film material, or not, depending on the surface chemistry established (Fig. 5.2(c)). The result could be the same as in the case of top-down process despite different order in which various operations were performed.

A Self-Assembled Monolayer (SAM) process is an example of the bottom-up patterning mechanism often implemented in various bio-engineering and other applications. Process of self-assembly in general, is the process where independent entities interact in the coordinated fashion to produce larger, ordered structures, featuring desired shape.

The point regarding this last issue is that all growth and shape defining processes occurring in nature are in their essence based on the bottom-up principle. An example is the pre-programmed by genetics growth and "patterning" of the humans beginning at the embryo stage.

5.2.3 *Lift-Off Patterning*

In the patterning process known as a lift-off, similarly to the bottom-up sequence, the deposition of the film to be patterned takes place after pattern has been defined in the layer of photoresist (Fig. 5.3(a) and (b)). Other than that, the essence of the lift-off process is rooted in the top-down scheme discussed earlier.

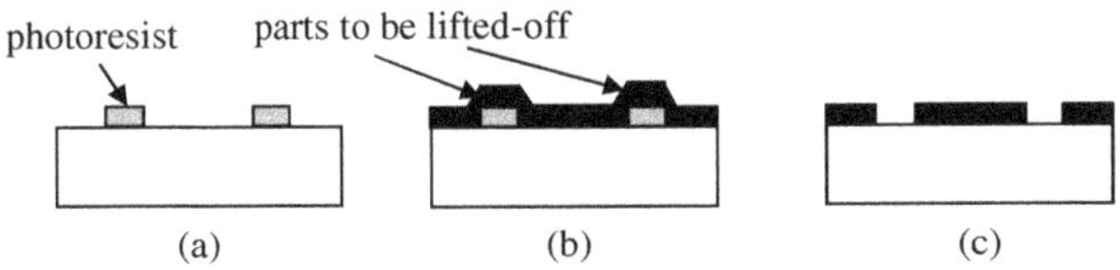

Fig. 5.3. Schematic illustration of the lift off process, (a) deposition and patterning of photoresist, (b) deposition of material to be patterned, and (c) dissolution of the photoresist resulting in material on top of photoresist to be lifted off and removed.

The lift-off is used with materials which because of the resistance to etching, for instance, gold, cannot be patterned following conventional top-down procedure. In the lift-off sequence the negative of the desired pattern is first defined in the layer of photoresist deposited on the surface (Fig. 5.3(a)). Then, a thin film of gold to be patterned is blanket deposited on the surface of the wafer as shown in Fig. 5.3(b). The pattern is transferred to the layer of gold by dissolving photoresist in the organic solvent and removing gold in the parts where it was covering photoresist and establishing the desired pattern in the thin-film gold remaining on the surface as a result (Fig. 5.3(c)).

5.2.4 *Mechanical Mask*

Unlike in the case of top-down and bottom-up processes in which thin-film deposition and patterning are applied in either deposition-patterning or patterning-deposition sequence, there are processes in which deposition and patterning occur simultaneously. The most common is the process using a thin sheet of metal or plastic with properly shaped openings cut into it and acting as a stencil during deposition process (Fig. 5.4(a)). The result is a deposited thin film featuring desired pattern (Fig. 5.4(b)).

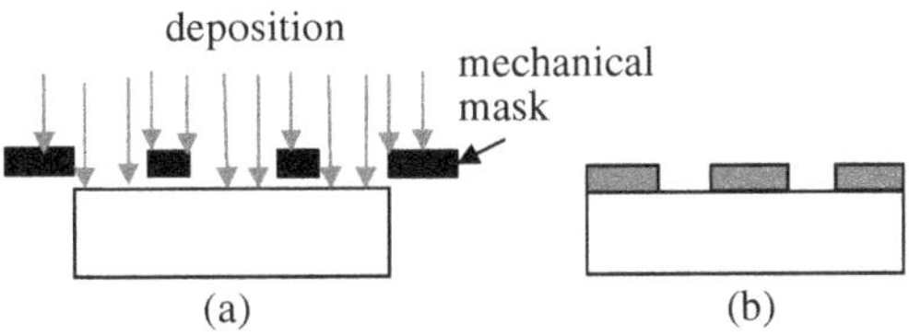

Fig. 5.4. Thin-film patterning using mechanical mask, (a) mask is blocking deposition in selected areas, (b) deposited material patterned using mechanical mask.

The use of mechanical masks, also known as shadow masks, is compatible only with selected deposition techniques discussed later in this chapter. Due to the resolution of pattern definition using shadow masks limited to micrometer regime and difficulties with precise alignment of multi-layer patterns defined using this technique, mechanical mask pattern definition is rarely used in mass industrial manufacturing of semiconductor devices. On the other hand, however, this patterning mode is often exploited in research and development laboratories, as well as in small scale production, taking advantage of low cost of the masks and overall simplicity of the process.

5.2.5 *Printing*

The concept of using printing and stamping to create 2D patterns, letters on paper, for instance, is known and widely exploited for centuries. In various forms and shapes it has been extended to a range of manufacturing processes including fabrication of semiconductor structures. In contrast to typical top-down and bottom-up processes, in printing and stamping operations of deposition and patterning occur simultaneously.

Discussing printing employed in commercial applications, a distinction needs to be made between 2D printing and 3D printing as they serve different purposes. While the former creates thin film, 2D patterns, the latter is essentially a technique which dispenses material such that 3D stand-alone structures in size determined by the size of the printing machine and printing precision are created. The 3D printing offers manufacturing capabilities unmatched by any other technique, and thus, its use is spread across the industries. In semiconductor manufacturing 3D printing is used, for instance, to print packages housing integrated circuits chips, or elements of MEMS devices, but is not used in the formation of ultra-thin nanoscale patterns upon which manufacture of the advanced semiconductor devices and circuits is based.

The 2D printing can be implemented using ink-jet printing technique which finds various unique uses in semiconductor technology including processes on flexible substrates in roll-to-roll (R2R) operations. Ink-jet printing is a technique commonly used in printers common in our homes and offices. In the conventional printing there is no need for the definition of very narrow lines because human eye cannot resolve the patterns much smaller than some 100 μm. The same printing technology adapted to the needs of nanoprinting can define patterns in the fraction of micrometer range.

In summary of the overview of various patterning schemes involved in semiconductor device manufacturing, the dominant role of the top-down patterning processes in the mainstream semiconductor devices and circuits manufacturing is recognized. Reflecting this state-of-the-art, further discussion of the patterning technology in semiconductor manufacturing processes is focused on the top-down scheme.

5.3 Processing Steps in Top-Down Sequence

Recognizing a range of pattern definition techniques available and used in semiconductor device manufacturing depending on the needs, further discussion in this chapter will be limited to the top-down pattern definition scheme performed on the rigid substrate wafers. The top-down sequence remains dominant across the range of device manufacturing scenarios and as such, adequately represents the mainstream semiconductor device technology.

As discussed earlier, a *p–n* junction diode well represents common semiconductor device configuration. Beginning with the structure resulting from the pattern defining sequence in Fig. 5.1, remaining operations involved in the *p–n* junction diode fabrication process are illustrated in Fig. 5.5. First, using selective doping process discussed later in this chapter, a portion of the *n*-type silicon wafer defined by the opening in the film of silicon oxide acting as a mask during the doping process is converted into *p*-type silicon forming a *p–n* junction (Fig. 5.5(b)). Next, wafer is covered with a thin film of metal (Fig. 5.5(c)) which is then shaped into contact following patterning sequence illustrated in Fig. 5.1 to form an ohmic contact to the *p*-type region. In the separate deposition step thin

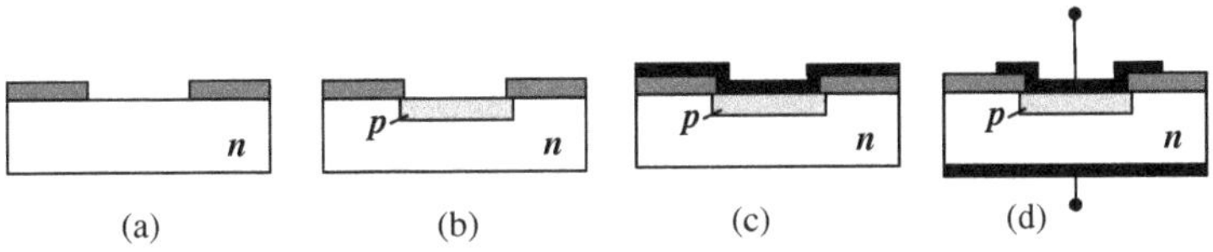

Fig. 5.5. Steps in *p–n* junction fabrication sequence include (a) patterning of the oxide, (b) *p*-type doping forming *p–n* junction, (c) deposition of thin-film metal contact, and (d) patterning of the metal contact.

metal ohmic contact is formed on the back surface of the wafer to complete diode's fabrication process (Fig. 5.5(d)).

Basically, the same as shown in Fig. 5.5, or very similar sequence is used and reproduced as many times as needed in any typical semiconductor device manufacturing process. The difference is that in the case of the manufacture of a simple discrete device, such as *p–n* junction diode in Fig. 5.5, the process may involve just two patterning steps, and few dozen or so operations performed on the wafer. In the case of the complex integrated circuit manufacturing, however, over twenty patterning steps and hundreds of operations performed on the wafer may be needed to complete the process.

Using as a guideline the sequence started with Fig. 5.1 and completed in Fig. 5.5, the following key types of operations performed on the wafer in the commonly employed top-down manufacturing sequence are listed below, and then methods of their implementation schematically summarized in Fig. 5.6.

- Surface processing including surface cleaning.
- Thin-film deposition which represents broadly defined additive processes.
- Pattern definition using methods of lithography.
- Subtractive processes which are used to remove material by etching.
- Selective doping which introduces dopant atoms into semiconductor material.
- Processing contacts and interconnects including surface planarization.
- Assembly and packaging which convert device existing as a part of the wafer into device which can be used to build electronic circuit.

The same in nature operations performed at the different stages of the complex device/circuit manufacturing sequence are typically carried out in somewhat modified fashion. For instance, surface cleaning prior to gate dielectric deposition is subject to different requirements than surface cleaning operations carried out before contact metallization. Taking it into consideration, processes involved in especially elaborate fabrication sequences are divided into front-end-of-line (FEOL) and back-end-of-line (BEOL) processes with the first metallization step in the sequence being, somewhat arbitrarily, adopted boundary between these two processing modes.

In the remaining sections in this chapter each type of operations listed above is considered by reviewing methods used to implement them. As an

Cleaning/Surface preparation	▪ Liquid phase (wet) ▪ Gas phase (dry) ▪ Supercritical	
Additive Processes	▪ Oxidation ▪ Chemical Vapor Deposition (CVD) ▪ Physical Vapor Deposition (PVD) ▪ Physical Liquid Deposition (PLD) ▪ ElectroChemical Deposition (ECD)	CVD, PVD → Epitaxial deposition → Non-epitaxial deposition
Pattern definition	▪ Optical (photo) lithography ▪ E-beam lithography ▪ X-ray lithography	
Subtractive Processes	▪ Liquid phase (wet) etching ▪ Gas phase (dry) etching	
Dopant Introduction	▪ Diffusion ▪ Ion implantation	
Planarization/wafer thinning	▪ Chemical-Mechanical Planarization ▪ Scrubbing	

Fig. 5.6. Types of operations performed on semiconductor wafer in the manufacture of semiconductor device before assembly and packaging, and methods of their implementation.

introduction to this review Fig. 5.6 identifies types of operations and corresponding techniques used to implement them.

5.4 Surface Processing

As was pointed out earlier in this volume, conditions of the surface of the substrate as well as conditions of the surfaces of the films formed on the substrate have a critical impact on the outcome of the process in terms of device performance and manufacturing yield. It is therefore imperative that contamination-free surfaces of the processed wafers are maintained throughout the entire manufacturing sequence (Ruzyllo, 2025).

This section considers selected key issues related to surface processing technology in semiconductor device manufacturing. First, surface cleaning processes are considered. Then, merits and implementation of operations referred to as surface conditioning are discussed.

5.4.1 *Surface Cleaning*

To reiterate points made in Section 4.6 of this volume, any surface contaminant, i.e. element other than the host material, or particulate matter located on the surface of semiconductor wafer, will have an adverse, disruptive effect on any operations performed on this surface. In turn, any contamination related process malfunction will ultimately lead to the malfunction of devices formed on the contaminated surface. Therefore, surface processing step aimed at the removal of contaminants needs to be performed numerous times at various stages of the manufacturing process. In fact, in the case of ultra-high packing density silicon integrated circuits, wafer cleaning is the most frequently applied processing step. On the other hand, cleaning processes play a somewhat less critical role in the case of materials and materials systems in which inherently inferior material characteristics rather than features of the surface determine the performance of the device manufactured.

Surface cleaning is the process aimed at the removal of solid, inherently non-volatile contaminants such as particles (Vali, 2009) or metallic impurities from the surface without uncontrolled alterations of its characteristics (Meuris *et al.*, 1994). Figure 5.7 schematically illustrates possible mechanisms of surface cleaning in semiconductor device manufacturing. First one involves chemical reaction between appropriately selected chemically reactive species and surface contaminants forming soluble in liquid ambient or volatile in the gas-phase compounds used to remove contaminants from the surface (Fig. 5.7(a)). Very often the process of dislodging contaminants from the surface is enhanced by the physical motion of the ambient as shown in Fig. 5.7(b).

Other than that, chemically neutral physical species carrying out kinetic energy can be used to knock contaminants, for instance, particles, from the surface of the wafer (Fig. 5.7(c)). Yet another option involves exposure of the wafer to the UV or IR (infrared) light causing desorption of the contaminant from the surface (Fig. 5.7(d)). Different considerations apply to the cases where surfaces subject to cleaning involve deep 3D features (Bakhtari *et al.*, 2006).

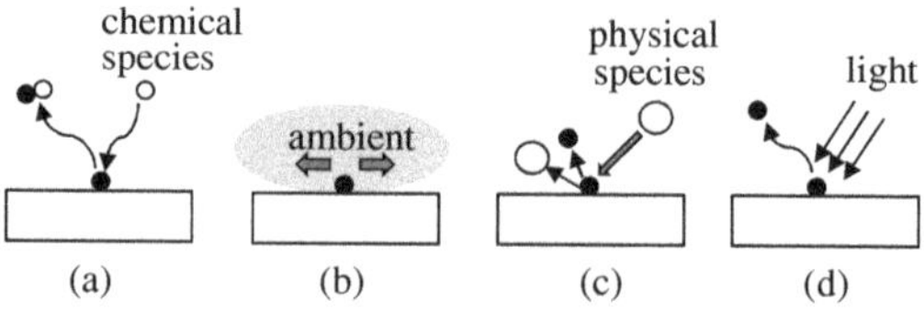

Fig. 5.7. Implementation of the surface cleaning processes in manufacturing of semiconductor devices.

Operations shown in Fig. 5.7 can be implemented either in the liquid-phase (wet cleaning), or in gas-phase (dry cleaning), (Ruzyllo, 2014) both of which are considered below. In addition, the following discussion touches on the supercritical cleaning technique which represents somewhat different mode than wet and dry cleanings employed in semiconductor fabrication. As a reminder, characteristics of the wet and dry processes in semiconductor device technology were discussed in Sections 4.3 and 4.4, respectively.

Wet cleaning: During the most commonly used in semiconductor device fabrication wet cleaning operations, contaminants are removed via selective chemical reactions in the liquid-phase which causes either their dissolution in the solvent, or conversion into the soluble compounds (Fig. 5.7(a)). Typically, process is enhanced by physical interactions with a liquid ambient by means of megasonic agitation (Fig. 5.7(b)), or, in extreme cases such as post-polishing/planarization cleaning, scrubbing with a soft brush. Any process involving physical interactions of this nature needs to be carried out very carefully so that collapse of the patterns existing on the processed surface, and other physical damage to the surface features, is prevented.

Specificity of wet cleaning is such that there are no single chemistries that would be equally effective in the removal of all types of contaminants. Therefore, cleaning procedure known as RCA cleaning or Standard Cleaning (SC) comprise of multi-step sequences each involving differently formulated cleaning solutions (Kern and Puotinen, 1970). For instance, in the case of silicon processing, particles removal step typically involves NH_4OH:H_2O_2:H_2O mixture (ammonium hydroxide-hydrogen peroxide mixture, or APM), also known as SC1 (Standard Clean 1), or RCA1. To prevent roughening of Si surface during particle removal process (Knotter *et al.*, 2000), weak APM solutions, (e.g. 1:1:50, at temperature of 40 °C) are used, typically in conjunction with megasonic agitation. To remove heavy organic contamination, strongly oxidizing H_2SO_4(4):H_2O_2(1) solution at 100–130 °C, also known as SPM (Sulfuric Peroxide Mixture), or "piranha clean" is commonly used. The same goal can be accomplished by using sulfuric acid-ozone mixture (SOM). The spontaneously formed on the silicon surface during cleaning an ultra-thin layer of silicon oxide can be readily removed by means of a brief exposure to diluted HF:H_2O (DHF) solution at the ratio of 1:100 or weaker.

The complete wet cleaning processes are typically implemented by sequential immersion of processed wafers, typically in batches, in the properly formulated cleaning solutions contained in the separate tanks installed in the robot-operated tools known as wet benches. Some tanks in the tool are fitted with transducers sending megasonic waves across the liquid in the direction parallel to the wafer surface (megasonic agitation) (Mayer, 1979; Mouche and Tardiff, 1995). Each cleaning step implemented in the wet bench is followed by the DI water rinse while the entire sequence is concluded with a wafer drying process as discussed in Section 4.3.3.

When it comes to wet cleaning of compound semiconductors, composition of wet cleaning solutions needs to be adjusted according to the chemical makeup of the material, although, discussed above general rules applicable to wet cleaning procedures remain the same.

A special role in wet wafer cleaning in semiconductor technology is played by the contamination removal operation known as post-CMP cleaning. Referred to as scrubbing, it is used to remove heavy contaminants from the wafer surface by means of mechanical interactions implemented in the form of brush scrubbing, or megasonic scrubbing. This heavy-duty cleaning mode is employed primarily after the Chemical-Mechanical Planarization (CMP) operations (Oliver, 2004).

Dry cleaning: An alternative to the standard wet cleans using large amounts of liquid chemicals and deionized water is a dry cleaning technology (Heyns *et al.*, 1999). In this case removal of contaminant from the surface takes place via chemical reaction in the gas-phase converting it into a volatile compound (Fig. 5.7(a)), or as a result of momentum transfer between species impinging on the surface and surface contaminants (Fig. 5.7(c)), or as a result of surface irradiation (IR–heating, UV–bond breaking/oxidation) sufficient to overcome forces causing volatile contaminant to adhere to the surface (Fig. 5.7(d)). In the case of processes carried out at the reduced pressure, plasma can be used as an effective dry cleaning enhancing agent. For instance, hydrogen-containing plasmas are used to remove fluorocarbons or oxides disturbing processed surfaces. Furthermore, the gas-phase equivalent of HF:water process used to reduce oxide on silicon surface is water-free anhydrous HF (AHF) mixed with the vapor of the alcoholic solvent such as methanol, or ethanol (Torek *et al.*, 1985).

Due to the inferior to wet cleans "cleaning strength", dry cleaning methods are used primarily to control chemical composition of semiconductor surface rather than to remove contaminants such as particles and metallics from the grossly contaminated surfaces. The former can be removed in the gas phase using cryogenic cleaning, but not with 100% efficiency across all particle sizes and not always effectively enough from the non-planar surfaces. Volatilization of metallic contaminants using gas-phase chemistry such as, for instance, UV-promoted chlorine (Cl_2) exposure is typically accompanied by the roughening of the processed surface due to chlorine reaction with silicon.

Supercritical cleaning: To be effective, cleaning medium either wet or dry must reach contaminated areas on the wafer surface and then remove products of the cleaning reactions from the surface. The complex features formed on the surface, such as high aspect ratio trenches, for instance, cannot be penetrated using liquids because of the surface tension. At the same time methods using dry cleaning media cannot dislodge particles from such deep features. To work around these limitations supercritical fluids (SCF) need to be used to carry cleaning chemistries into the trench and then wash products of the cleaning reaction out of the trench. At certain pressure and temperature (critical point) either gas or liquid can be transformed into a supercritical fluid (SCF) (Clifford, 1998) which by combining some properties of liquids and gases displays distinct properties. With density not much smaller than density of the typical liquid, and viscosity comparable with gas, supercritical fluid is well suited for cleaning wafers with ultra-small geometries. As SCF features negligible surface tension, a complete penetration of high aspect ratio structures can be accomplished as opposed to water-based cleaning solutions.

5.4.2 *Surface Conditioning*

With process gases, process chemicals, and overall process ambient getting gradually cleaner on one hand, and the high cost of high purity chemicals and water on the other, gradual reduction of cleaning operations in semiconductor device manufacturing appears to be established trend. Instead, increased emphasis is placed on surface conditioning processes (Ruzyllo, 2025) which are designed to enforce desired chemical composition of the surface of the processed semiconductor substrate whether locally (see surface functionalization in Fig. 5.2), or globally (surface of the entire substrate).

Surface conditioning operations can be carried out using either wet or dry chemistries as well as tools and methods used in surface cleaning operations. The goal is to enforce surface termination assuring stable, time and ambient resistant, as well as reproducible surface characteristics. The essence of the of the process illustrates Fig. 5.8, where initially unstable silicon surface covered with residual oxide and hydrocarbon species resulting from the uncontrolled interactions with the ambient air (Fig. 5.8(a)) is replaced through the series of operations ending with HF:H_2O treatment with hydrogen termination (Fig. 5.8(b)) creating chemically stable silicon surface. Somewhat less complete hydrogen termination of Si surface can be obtained using mentioned earlier anhydrous HF (AHF) mixed with the vapor of water or alcoholic solvent.

As it could be expected, the way surface is terminated has an effect on surface energy. Changes in the surface energy in turn, change the way any given surface interacts with water. In general, surfaces which lend themselves to wetting are known as hydrophilic surfaces. Those which repel water are known as hydrophobic surface. In the case of silicon, surface shown in Fig. 5.8(a) displays hydrophilic characteristics while hydrogen terminated surface (Fig. 5.8(b)) features strong hydrophobic characteristics. Also, chemically pure Si surface features hydrophobic characteristics. In contrast, surface covered with residual oxide and organic contaminants features hydrophilic characteristics and is unstable when exposed to the environment.

In general, carefully executed surface conditioning operations are needed the most prior to some key deposition steps such as those involving formation of the epitaxial layers, MOSFETs gate dielectrics, or ohmic contacts. Special requirements regarding the performance of surface processing are imposed on the treatments applied prior to high-temperature

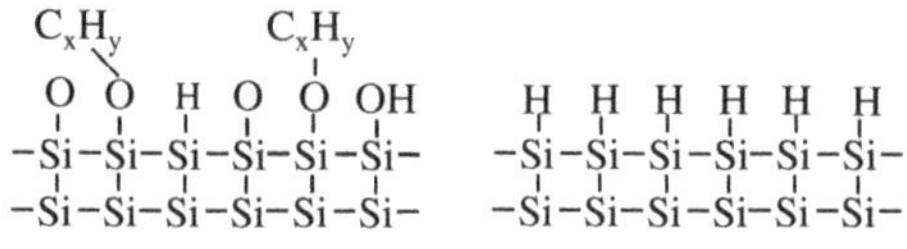

(a) (b)

Fig. 5.8. (a) Silicon surface as exposed to ambient air and (b) hydrogen terminated Si surface.

deposition steps as temperature promotes transformation of surface deficiencies into permanent defects.

5.5 Additive Processes

As indicated on several occasions in the previous chapters, semiconductor devices include multilayer structures comprised of various materials (semiconductors, insulators, metals) processed in the form of thin films. Referred to as additive processes, are those which add material in the form of thin film on top of the substrate where upon adequate patterning and equipped with electrical contacts, resulting multi-layer material system acts a functional semiconductor device (Smith, 1995). Considering the role they are fulfilling, additive processes are at the core of semiconductor device manufacturing technology.

5.5.1 *Characteristics of Additive Processes*

The first consideration in the process of thin-film deposition is a crystal structure of the deposited film. If it needs to be in the single-crystal form, then according to the rules of epitaxial deposition discussed in Chapter 2 the substrate upon which film is deposited needs to be a single-crystal solid, crystallographic structure of which will be reproduced in the deposited film (Fig. 2.13(a)). As discussed in Section 2.7, epitaxial deposition processes constitute a separate class of additive processes requiring special attention in terms of the substrate quality and process control. Thin films of amorphous and polycrystalline materials can be deposited on any substrate assuming it is compatible with the conditions of the deposition process used (Sesham, 2012).

An important consideration in the discussion of additive processes is the extent to which substrate material is involved in the chemical reactions forming thin film. In this respect a special case of thermal oxidation of silicon and growth of its native oxide (silicon dioxide SiO_2) is a result of chemical reaction at the elevated temperature in which silicon atoms from the substrate wafer react with oxygen in the process ambient. Very thin part of the silicon surface region is "consumed" in the process and the gradual transition from the silicon composition to SiO_2 composition results in the smooth chemical interface in the Si-SiO_2 structure. The same

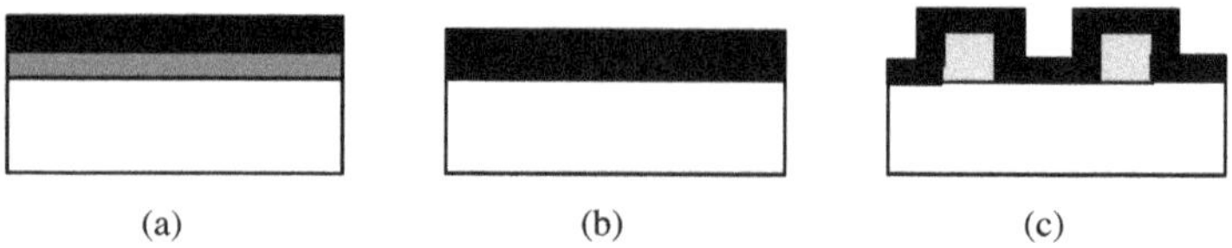

Fig. 5.9. (a) Chemical and structural interface (transition) between thin film and the substrate, (b) abrupt interface, (c) conformal coating.

structure also features structural interface within which single-crystal structure of Si transitions into amorphous structure of SiO_2 (Fig. 5.9(a)). The process of thermal oxidation of silicon and characteristics of the Si-SiO_2 structure are further considered in the following section.

In the case of the most broadly used additive techniques, substrate does not participate chemically in the film formation, which means that all the components of the deposited film are supplied from outside of the substrate wafer. In such cases deposition process does little to improve the characteristics of the substrate wafer and transition from the substrate to the thin film is featured by abrupt change of material composition at the interface, or in other words, abrupt chemical interface (Fig. 5.9(b)). If the substrate and thin-film materials feature in addition different crystallographic structure, the material system formed features also the abrupt structural interface

Regardless of the crystallographic structure of the deposited material, growth mechanism, and deposition technique, the process and its outcome must meet certain requirements. First, the film must be homogeneous in terms of its chemical composition and crystallographic structure. Any departure from material homogeneity needs to be considered as a defect potentially causing device malfunction. Second, precise control over the film thickness during the deposition process and uniformity of thickness of deposited film over the entire area coated must be assured. Furthermore, uniformity of film thickness understood as a conformality of coating over the non-planar features of the surface of the processed substrate is a stringent requirement in some deposition processes (conformal coating in Fig. 5.9(c)). Finally, deposition process must be compatible with the type of substrate used. For instance, additive processes requiring elevated temperature cannot be used in conjunction with glass, or plastic substrates. It is with these requirements and constraints in mind suitability of any given deposition method needs to be considered.

5.5.2 *Thin-Film Growth by Oxidation: Thermal Oxidation of Silicon*

When the thin film to be formed on the surface is an oxide of the substrate material, then the process of the film growth by oxidation of the latter can be considered. In general, when external conditions allow surface atoms of the solid substrate react with oxygen or oxygen containing species in the ambient, and nucleation of the native oxide of the substrate material results. Depending on the material and external conditions, it may continue toward formation of the mechanically coherent oxide film. The process is common in nature and in our surroundings and is typically associated with material deterioration. In the case of metals, it is known as material degrading process of corrosion.

Among semiconductor materials, only silicon forms on its surface native oxide in the form of silicon dioxide SiO_2 which features device-grade characteristics. Among compound semiconductors, only silicon carbide (SiC) can get oxidized to form functional oxide which is also a silicon dioxide SiO_2 as during the process of oxidation gaseous carbon oxides to not get incorporated into the oxide formed on SiC surface.

Growth of oxide on the silicon surface can be promoted in various ways (Hoff, 1988). Using methods employed in semiconductor manufacturing it can be an oxidation process stimulated by oxygen plasma (plasma oxidation), or oxidation through electrochemical reaction in the liquid electrolyte (anodic oxidation). In silicon device manufacturing the most common process is using thermal energy to promote oxidation of silicon. The process of thermal oxidation of silicon is briefly considered as follows.

The chemical reaction driving thermal oxidation of silicon in oxygen is $Si + O_2 \rightarrow SiO_2$ and is referred to as dry oxidation. When instead of dry oxygen water vapor is used as an oxidizing agent the oxidation proceeds according to the reaction $Si + 2H_2O \rightarrow SiO_2 + 2H_2$ and is known as wet oxidation. While rate of oxidation using water vapor is faster than in the case of dry oxidation, the oxidation kinetics qualitatively remain the same. In both these cases, oxidation reaction occurs at the silicon surface which means that at the early stage of oxidation, when there is not oxide on the surface or it is very thin, oxidizing species have direct access to the surface and the oxide grows at the high rate which is determined by the surface reaction rate dependent on the composition of near-surface region of silicon (Ridley *et al.*, 1995). Later in the process oxidizing species need to diffuse across the oxide already formed to reach Si-SiO_2 interface where oxidation reaction is taking place and the diffusion-controlled oxidation process slows down (Hossain *et al.*, 1990).

The two stages of thermal oxidation of silicon described above are reflected in the oxide growth kinetics and can be qualitatively illustrated by means of the oxide thickness x_{ox} vs. time of oxidation t relationship as shown in Fig. 5.10. During an early surface reaction-controlled stage of oxidation, x_{ox} vs. t relationship is linear and the process is defined as linear growth. As the oxide grows thicker, oxidation gradually transitions to the slower parabolic growth controlled by the diffusion of the oxidizing species where x_{ox} vs. t relationship becomes parabolic.

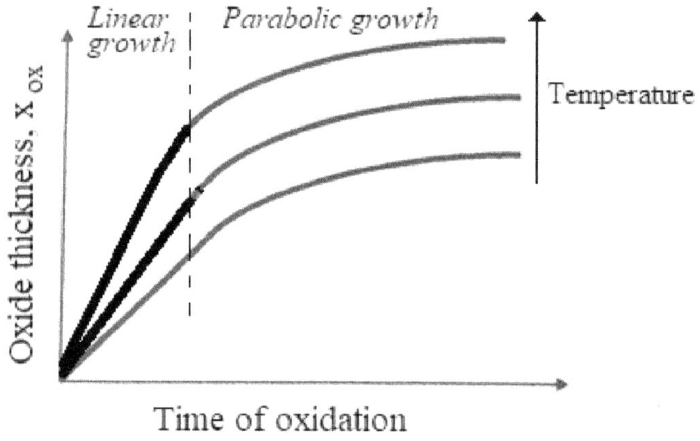

Fig. 5.10. Kinetics of the thermal oxidation of silicon.

As Fig. 5.10 also shows, higher temperature of oxidation results in the faster rate of oxidation and thicker oxide for the same time of oxidation. As mentioned earlier, significant increase of the thermal oxide growth rate, however, at the expense of oxide integrity, is accomplished by using water vapor containing ambient (wet oxidation) instead of dry oxygen (dry oxidation). Also, growth rate of thermal oxide depends on the surface orientation of the single-crystal substrate wafer and in the case of silicon is faster for (111) oriented surface than for (100) oriented surface.

In silicon device fabrication process of thermal oxidation is implemented using either horizontal or vertical furnaces shown in Fig. 4.6 equipped with adequate gas delivery infrastructure. Temperature of silicon oxidation used in furnace processing varies from about 700 °C to about 1,000 °C. In the processes requiring low-thermal budget, the RTP tools can be used to implement the process known as rapid thermal oxidation (RTO) (Fig. 4.7). In either case, the additional level of control over the oxidation rate is available by controlling partial pressure of oxygen inside the oxidation chamber. Slower oxidation rate at any given temperature can be accomplished by mixing oxygen with nitrogen, and thus, reducing oxygen partial pressure in the oxidation chamber. On the other hand, by increasing oxygen pressure above atmospheric (high-pressure oxidation, HIPOX), significant increase of the oxidation rate at any given temperature can be enforced.

In the process runs where the growth of the oxide featuring desired thickness would require high temperature damaging to the processed structure, reduction of oxidation temperature can be accomplished by employing plasma enhancement of the process. For instance, by using remote-plasma oxidation (see remote plasma apparatus in Fig. 4.9) which combines plasma enhancement of the process with thermal oxidation, temperature of the wafers subject to oxidation can be significantly lower than in would be the case of thermal only oxidation only.

In silicon technology thermal oxidation is among common processes used. When and how it is being used depends on the purpose thermally grown oxide is meant to serve. Its main use is in the growth of gate oxide in MOS/CMOS transistors where advantage is taken of the high-quality Si-SiO_2 interface conducive with the needs of the MOSFET operation. Depending on application of transistor, thermal gate oxide thickness varies from few nanometers in the case of the MOSFETs comprising logic integrated circuits, to about 50 nm in the case of the power MOSFETs. In the former case, as earlier discussion indicated, if the transistors design is requiring gate oxide thinner than some 1.5 nm, then the SiO_2 is replaced as a gate oxide by the high-*k* dielectric material.

In addition, thermal oxidation of silicon is used to form oxide on its surface to protect the surface, isolate features formed on the surface, and to grow oxide to be used as a mask during selective doping processes (see Section 5.8). In such cases required oxide thickness may range from some 100 nm to 500 nm and above, which typically calls for the use of wet oxidation.

5.5.3 *Physical Vapor Deposition* (*PVD*)

This class of thin-film deposition methods is broadly used in various technical endeavors beyond semiconductor device manufacturing. As the name indicates, PVD processes in the purest form are based on the physical effects with no chemical reactions involvement at any stage of the film deposition process. What it means is that the source material is physically transferred in the form of vapor to the substrate where it forms thin-film without alteration of its chemical composition. An exception to this rule is a process of reactive physical vapor deposition where before condensing on the substrate vapor is subject to chemical reactions with gases separately introduced into the process chamber and modifying chemical composition of the deposited film.

To execute physical transfer of the matter from the solid source to substrate, source material needs to be volatilized and then transported from the source to the substrate in the gas phase. To prevent any disturbance of the transfer process, as well as to promote volatilization of the material to be deposited, the entire process must be carried out in vacuum. Thus, all PVD processes are inherently carried out in the high-vacuum chambers.

The PVD techniques can be distinguished based on the method used to volatilize the material which is then deposited on the substrate in the form of thin film. Schematics in Fig. 5.11 explain principles of three different ways physical vapor deposition is implemented. With a solid-state source material in each case, the first technique is using heat to melt a solid and to cause its evaporation (Fig. 5.11(a)). The vapor is moving in vacuum toward the substrate where it solidifies forming a thin film. An alternative way of PVD process implementation is based on the discussion in the previous chapter sputtering process (Fig. 5.11(b)).

The three techniques identified in Fig. 5.11 and referred to as thermal evaporation, sputtering, and molecular beam deposition. The last

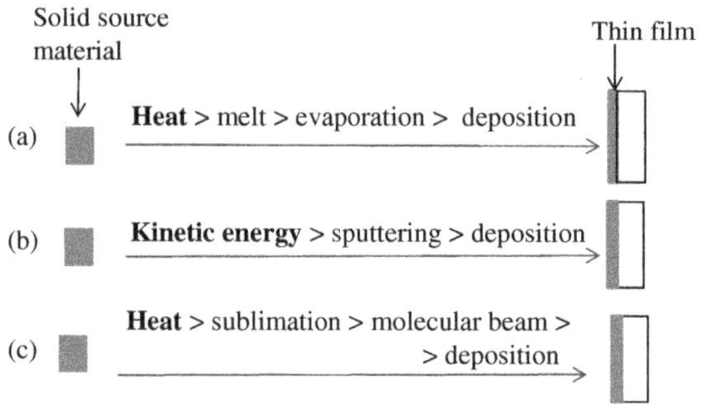

Fig. 5.11. Processes involved in the physical vapor deposition including (a) thermal evaporation, (b) sputtering, and (c) molecular beam.

one (Fig. 5.11(c), in its most common version is known as molecular beam epitaxy (MBE) because of how it is predominantly used. Some other, more specialized PVD methods, such as, for instance, ion beam sputtering, or ion cluster beam deposition are not considered in this introductory discussion.

Thermal evaporation: Deposition of thin film by thermal evaporation requires at first melting of the source material after which its vaporization takes place. Evaporated species will then condense on the surface of the substrate exposed to the vapor coating it with film of the source material. To make undisturbed transfer of material in the vapor phase possible, any residual gases from the reaction must be evacuated to obtain in the deposition chamber vacuum of at least 10^{-6} Torr. Under such conditions, the vapor moves undisrupted from the source to the substrate.

Schematic diagram of the reactor used for thermal evaporation is shown in Fig. 5.12(a). It consists of a bell jar connected to vacuum pumps. Unlike in the simplified for the sake of illustration arrangement shown in Fig. 5.12(a), in commercial tools substrate wafers are mounted inside the bell jar on the dome-shaped (planetary) fixture rotated with respect to the source during the evaporation process to assure its wafer-to-wafer uniformity.

A distinguished element in thermal evaporation process is a heating technique used to melt the source material. Two methods used for this purpose include resistance heating, and heating by an electron beam (*e*-beam). In the first case, the material to be evaporated remains in

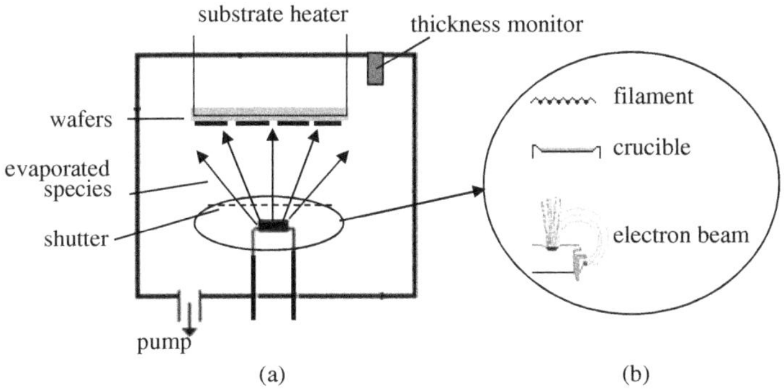

Fig. 5.12. Process of thermal evaporation, (a) reactor and (b) filament, crucible, and *e*-beam evaporation modes.

contact with appropriately shaped refractory metal, for instance, tungsten in the form of the boat (crucible), or coiled wire (filament) through which high current is passed (Fig. 5.12(b)).

An alternative solution is offered by the electron beam heating of the source material. The high current and high energy *e*-beam is generated and then directed by a magnetic field towards the source (Fig. 5.13). The electrons striking the surface of source material causes its melting, then evaporation. As melting is localized and the molten part of the source material is not in contact with a crucible, contamination of the melt with species leaching out of the crucible is avoided. As a result, thin films formed by *e*-beam evaporation are less contaminated than the filament or crucible evaporated counterparts.

Thin-film deposition by vacuum thermal evaporation is limited to materials featuring relatively low melting point such as metals, including gold and aluminum, for instance. Also, discussed earlier small-molecule organic semiconductors are typically deposited on glass or on flexible substrates by thermal evaporation.

Sputter deposition: Deposition of higher melting point materials by means of physical vapor deposition requires other than thermal evaporation technique to volatilize solids in high vacuum. The solution is offered by the effect of sputtering discussed earlier (Fig. 4.10(b)). In the case considered here, sputtering is exploited in the process of sputter deposition.

Physical vapor deposition by sputtering uses plasma discharge in the space between the substrate and the source material to carry out thin-film deposition process. Plasma generated ions are accelerated by electric field toward the source material (referred to as a target), and upon imping on its surface cause ejection of atoms by momentum transfer. Atoms ejected

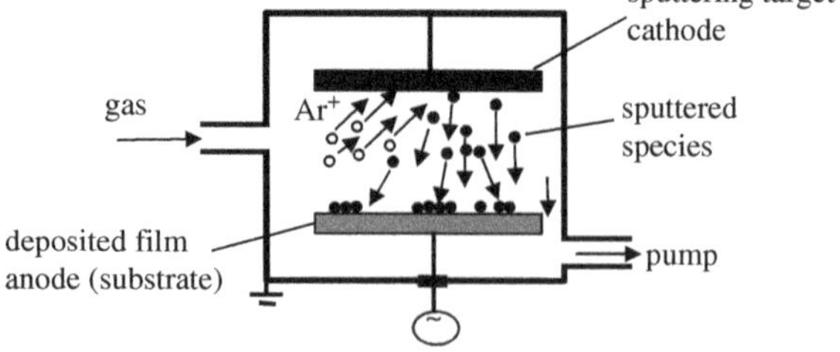

Fig. 5.13. Schematic illustration of the sputtering deposition process.

from the target are moving toward the substrate wafer where they adhere to the surface forming a thin solid film.

The sputter deposition process takes place in the vacuum chamber to which process gas, usually argon, is introduced to initiate plasma discharge (Fig. 5.13). Pressure of the process gas is maintained at the level from 5 to 20 millitorr needed to support plasma discharge. At this pressure, however, atoms ejected from the target are subject to numerous collisions before reaching the substrate. For that reason, the distance between the target, and the substrates during the sputter deposition process must be small. More than one target can be installed in the sputtering system allowing deposition of binary compound. To achieve appropriately uniform distribution of sputtered material, the total area of target materials should not be smaller than the total area of substrate wafers.

Among various powering schemes considered in Section 4.5.2, the AC power with a frequency in the radio frequency (RF) range, typically set at 13.56 MHz, is most used. Referred to as RF-sputtering, the process allows greatest versatility as it can be successfully used in the deposition of high melting point materials which are electrically conductive (metals) and non-conductive (insulators). If needed, chemical composition of the deposited material can be altered in the process of reactive sputtering.

A significant modification of the conventional sputter deposition tool involves addition of the permanent magnets to the cathode by the way of which a thin-film deposition scheme known as magnetron sputtering is implemented. The prime role of magnetic field created in the vicinity of the cathode is to confine electrons within the plasma discharge region. By doing so, the number of ionizing collisions between electrons and argon atoms increases leading to a higher rate of sputter deposition. Also, by deflecting electrons from their motion toward the substrate, undesired heating of the substrate by electrons impinging on its surface is prevented.

Overall, because of the high deposition rates, superior uniformity of deposited films, as well as limited, comparing to other sputtering methods, exposure of the substrate to the highly energetic, and thus potentially damaging plasma environment, makes magnetron sputtering broadly used sputter deposition technique.

Molecular beam epitaxy (MBE): Neither thermal evaporation nor sputtering allows thin-film deposition with atomic layer precision, and hence, neither is compatible with a needs of epitaxial deposition forming an ultra-thin film of single crystal on the lattice matched substrate (see discussion in Section 2.7). To achieve this level of control, species released from the source must be formed into a molecular beam. A molecular beam is formed by locally maintaining gas at higher pressure and then allowing its expansion through the small nozzle into a chamber maintained at much lower pressure. In the beam formed this way, particles (atoms or molecules) are moving with approximately same velocities mostly in parallel path with very few collisions. The instruments forming such beams are called effusion cells.

The method employing molecular beams for the purpose of epitaxial deposition is known as MBE. The MBE is among important tools of nanotechnology including its uses in advanced semiconductor engineering where it is used to deposit various materials (Asahi and Horikoshi, 2019; Chambers, 2025).

The schematic diagram of MBE reactor is shown in Fig. 5.14. It consists of ultra-high vacuum process chamber and pumping infrastructure supporting vacuum in the range of 10^{-8}–10^{-9} Torr which is needed to carry out molecular beam epitaxial deposition. In MBE processes the solid source material transitions directly to the gas phase by sublimation, or in other words, without passing through the intermediate liquid phase. The transition is confined within effusion cells from which particles in the gas-phase are released through the small orifices into the vacuum chamber where they form molecular beams (Fig. 5.14). The number of effusion cells installed on the tool depends on the number of elements involved in the formation of the epitaxial layers and dopants needed to define conductivity type of any given layer.

In the case considered in Fig. 5.14, three cells containing Ga, As, and Al allow epitaxial deposition on GaAs substrate of various compounds

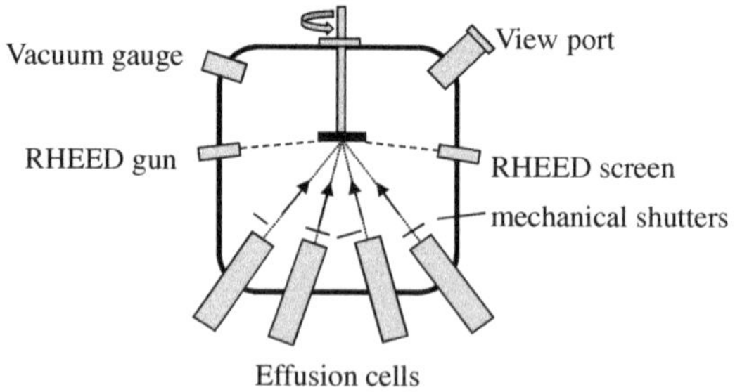

Fig. 5.14. Simplified schematics of the MBE process chamber.

within Al-Ga-As material system. During the MBE process abrupt changes in chemical composition of deposited films are accomplished by mechanical shutters which can instantaneously shut off the molecular flow of particles toward the substrate. Due to this feature, MBE allows formation of strained and relaxed superlattices, and quantum wells (see discussion in Chapter 2) comprised of the films down to the thickness below 1 nm which is basically the thickness of a single atomic layer. With this level of control over material composition and thickness, MBE allows precise bandgap engineering within multi-layer material systems.

During MBE deposition substrate is maintained at an elevated temperature allowing atoms arriving at the substrate surface to self-align with respect to the crystalline structure of the substrate and form undisturbed epitaxial layer. Depending on the processed materials, temperature of MBE deposition varies from about 400 °C to 900 °C. Elevated temperature in combination with ultra-high vacuum is also essential in the processes used to condition surface of the substrate before epitaxial deposition.

As an ultra-high vacuum process, MBE is compatible with a range of methods of physical analysis of solid surfaces which can only be implemented in the high-vacuum environment. The one often integrated with MBE tools for the purpose of in-situ characterization of growing epitaxial layers is reflection high energy electron diffraction (RHEED). As Fig. 5.14 shows, the RHEED instrumentation is installed such that growth of the film can be monitored based on the changes in the diffraction patterns produced by the electron beam impinging on the surface under the glancing angle.

The MBE deposition is used in the processing of the wide range of material systems from elemental silicon to quaternary III–V and II–VI multilayer structures wherever the highest precision epitaxial deposition process is required. Providing control of the crystal growth atomic layer by atomic layer, MBE allows tuning of electronic and photonic properties for particular device application. Resulting complex heterostructure geometries with precisely engineered bandgap make possible implementation of the range of electronic and photonic devices and circuits. In the former case it can be high-speed, low power-consumption transistors for logic ICs on one hand, and high-power transistors based on compound semiconductors on the other. In the latter case, the MBE process is employed in the development and manufacture of cutting-edge LEDs, laser diodes, and high-efficiency solar cells.

Due to the atomic-scale precision the MBE process is inherently relatively slow, and thus, features low manufacturing throughput. Therefore, while compatible with the fabrication needs of some highly specialized multi-layer structures listed above, the MBE method is not compatible with mass production of other types of devices requiring moderate temperature epitaxial deposition processes. In those instances, some methods among discussed in the following section, CVD methods could be viable alternatives.

5.5.4 *Chemical Vapor Deposition* (*CVD*)

In contrast to physical vapor deposition, where the material is deposited without changing its chemical composition, in CVD, deposited material is formed because of the chemical reaction in the gas-phase inside the process chamber (Choy, 2019). The reaction takes place on the surface of the substrate, or in its immediate vicinity. The reactants are supplied in the gas phase, and the chemical reactions involved are such that the deposited material is the only solid-state product. All byproducts of the reaction are in the gas phase, and hence, can be removed from the reaction chamber loaded with substrate wafers.

In principle, typical CVD processes are thermally driven. Depending on the chemistry used and composition of the substrate material, temperature of CVD processes may vary from as low as about 400 °C to as high as about 1,100 °C. For the purpose of illustration, two different types of CVD reactions requiring different temperatures are considered in Fig. 5.15. One type is based on thermal decomposition of the gaseous compound containing element to be deposited as a thin film (Fig. 5.15(a)). An example is thermal decomposition of silane gas SiH_4 into solid Si deposited on the surface in the form of thin film, and hydrogen which as

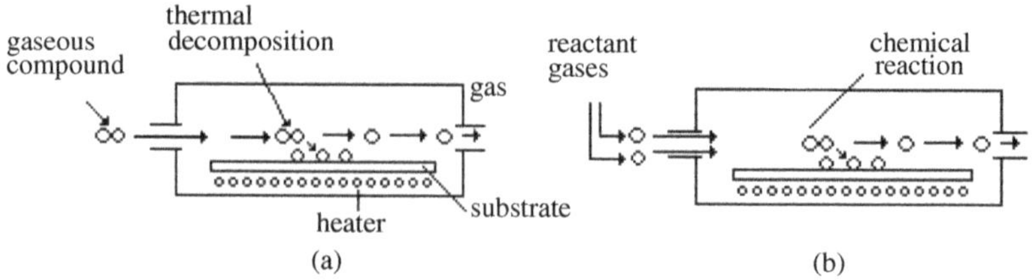

Fig. 5.15. CVD reactions (a) thermal decomposition, (b) chemical reaction between two gases.

a gaseous byproduct of the reaction is evacuated from the process chamber. The process of thermal decomposition of silane requires relatively high temperature of about 950 °C.

Alternative reaction path (Fig. 5.15(b)) represents the process in which two gases, for instance, silicon tetrachloride $SiCl_4$ and hydrogen H_2, are introduced into the process chamber. Silicon tetrachloride reacts with hydrogen at about 600 °C producing solid silicon forming thin film and gaseous HCl which is evacuated from the process chamber.

In most cases, CVD uses tools which are not very different from those used in other thermal processes. In practice, common CVD batch reactors are furnaces horizontal or vertical shown in Fig. 4.6 which by the virtue of being batch processors, offer the highest throughput. In some processes, in particular those in which CVD reactor is a part of the cluster tool, single-wafer CVD modules (Fig. 5.15) are used. Regardless of reactor geometry and number of wafers processed at a time, of importance are the dynamics of the gas flow inside the reactor. To produce uniform thin films, a laminar flow of gases inside the process chamber must be assured. Any turbulence in gas flow dynamics will cause localized changes in pressure and flow velocity of reactants resulting in the defective deposited films.

Depending on the needs, CVD processes can be implemented using various techniques introduced briefly as follows.

Low pressure CVD (LPCVD): The feature distinguishing various mainstream CVD schemes is the pressure at which process is carried out. The least demanding in terms of tool's complexity are CVD processes carried out at atmospheric pressure. While easiest to implement, Atmospheric Pressure CVD, commonly referred to as APCVD, produces films inferior in terms of the defect density to those deposited using CVD processes carried out at the reduced pressure, and known as Low Pressure CVD, or LPCVD in short. With pressure of reactants in the range form 0.1 Torr to 2 Torr and using the same as APCVD chemistry, LPCVD processes produce purer films featuring improved stoichiometry, lower defect density, better thickness uniformity, and superior step coverage. As a result, the LPCVD is very commonly used CVD method in manufacturing of semiconductor devices.

Plasma enhanced CVD (PECVD): In cases when CVD films are to be deposited on the surface covered with temperature sensitive materials, such as is the case in the back-end-of-line (BEOL) operations, the CVD processes need to be carried out at lower temperatures. In these instances, Plasma Enhanced CVD (PECVD) processes are being used. As discussed in Chapter 4, energy of electric field generating plasma supplements thermal energy in supporting CVD reactions which can now be carried out at the lower temperature.

The reactors using plasma enhancement of the process are similar regardless of whether plasma is used in additive processes such as sputtering and PECVD, or in subtractive processes such as etching discussed later in this chapter. What distinguishes various ways plasma is applied as a process medium in semiconductor manufacturing are the differences in the composition of gaseous chemistries, the way bias is applied, power used to generate plasma, and distribution of potential in the plasma.

Metalorganic CVD (MOCVD): Many III–V and II–VI semiconductor compounds thermally decompose at temperatures above 1,000 °C. Therefore, methods of epitaxial deposition operating at temperatures below 800 °C need to be available in compound semiconductor technology. Among CVD-based methods, MOCVD, is a technique allowing growth of the device-grade epitaxial layers at the temperature lower than those required by conventional CVD epitaxy, and pressures in the range from 10 to 760 Torr instead of ultra-high vacuum required by the discussed earlier MBE.

In semiconductor engineering, of interest are those metalorganic compounds which in addition to organic C_xH_y components contain elements of interest in a given semiconductor device fabrication. For instance, in the MOCVD process forming thin layer of single-crystal GaN, the source of gallium can be an organic compound $Ga(CH_3)_3$ while the source of nitrogen can be inorganic ammonia NH_3. The distinguishing features of MOCVD technology are lower temperature at which metalorganic compounds are thermally decomposed as compare to non-organic precursors used in conventional CVD processes and higher growth rates, and thus, higher throughput of MOCVD as compared to MBE.

Development of mass-production MOCVD technology was driven primarily by the needs of photonic devices fabricated using III–V compounds. Essentially all III–V and II–VI semiconductor compounds, and most of their alloys, can be deposited as high quality, uniformly doped single-crystal materials using MOCVD, regardless of whether in the single layer form or as multilayer complex heterostructures.

Overall, the MBE and MOCVD technologies play similar role of high-quality epitaxial layer deposition methods in III–V compound semiconductor device engineering. While the former successfully fulfils its function in smaller-scale commercial manufacturing, R&D, and prototyping, the latter is used in large-scale commercial manufacturing of light-emitting diodes, high-speed heterojunction transistors, and others.

Atomic layer deposition (ALD): Yet another variation of the chemical reaction driven, gas-phase thin-film deposition method, and as such, considered here to be a part of CVD class of methods, is Atomic Layer Deposition, ALD (Kaariainen *et al.*, 2013). Initially employed primarily as a technique depositing ultra-thin films of high-k gate dielectrics used in cutting-edge CMOS technology, it expanded into the method with a broader range of uses in semiconductor device engineering.

The difference between conventional CVD and ALD is that in the latter case deposited material is a result of two-stage chemical reaction applied in strictly executed sequence using two different gaseous precursors instead of single chemical reaction driving deposition of thin-film material in the former case. Figure 5.16 shows schematic representation of the ALD process which is typically carried out at temperatures not exceeding 300 °C and pressure in the mTorr to 1 Torr range.

Precursors A and B are introduced into the process chamber in the form of separated from each other pulses in sequence the A-B-A-B-A-B, etc. Precursor A adhering to and reacting with the properly conditioned substrate surface is needed to promote reaction with precursor B arriving at the substrate surface as a next pulse and completing formation of the material that is being deposited. The factor slowing down film growth by means of ALD is a required evacuation of all byproducts of chemical reactions following each reaction cycle. Referred to sometimes as molecular layering, the ALD process is self-limited because growth reaction stops when no adequate reaction sites are available on the surface. Therefore,

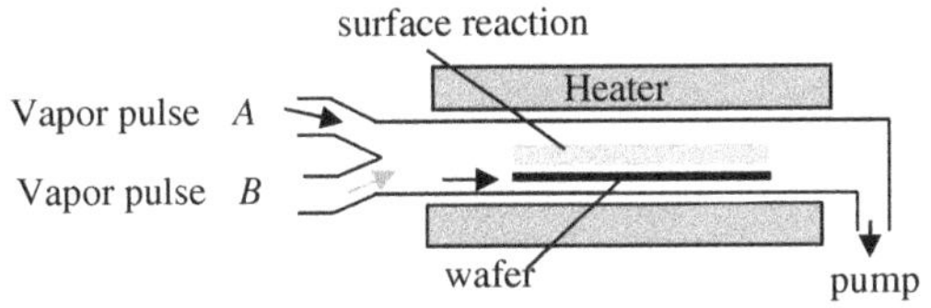

Fig. 5.16. Atomic Layer Deposition (ALD) process.

the ALD is particularly suitable for the precisely controlled layer-by-layer, highly conformal deposition of very thin films with sharply defined interfaces in the stacks of materials including oxides, metals, nitrides, metals, and others (Jackson *et al.*, 2011).

As the overview of CVD processes in this section indicate, this class of thin-film deposition methods offers unique combination of versatility, performance, and high process throughput. Using CVD, processes epitaxial deposition of single crystal as well as deposition of polycrystalline, and amorphous materials can be readily implemented. Essentially any semiconductor material, whether elemental or compound, can be deposited in the form of thin film by means of CVD. To a significant degree the same applies to thin-film insulators and metals. Furthermore, CVD technology offers a range of choices in terms of process pressure and temperature. Finally, thin films are deposited by means of CVD with adequate thickness control and feature good step coverage.

For all these reasons, the Chemical Vapor Deposition, CVD, in its many variations, is a key additive process in semiconductor device technology.

5.5.5 *Physical Liquid Deposition (PLD)*

In addition to solid-state precursors used in PVD processes and gaseous precursors used in CVD processes, liquid precursors are commonly used in semiconductor processing. Liquid source can be converted into gas which then will act as a reactant in chemical vapor deposition processes such as MOCVD. Alternatively, viscous liquid precursors can be physically applied to the wafer surface and then solidified by thermal curing. An element distinguishing various techniques in this area is the way in which viscous liquid precursors are applied to the substrate surface. Processes implementing PLD additive processes are reviewed as follows.

Spin-on deposition: The common method of physical liquid deposition (PLD) is a spin-on process, also known as spin coating, in which a controlled amount of liquid is dispensed onto the substrate surface and distributed over it by centrifugal forces created by wafer rotation at thousands of revolutions per minute. The thickness of the film is controlled by the substrate spinning rate (revolutions per minute, rpm) and the time of spinning. As a last step in the sequence, heating of the coated substrate in ambient air in the temperature range typically below 200 °C is applied to evaporate solvents and solidify deposited liquid into a solid film.

The spin-on process is simple to implement and is commonly used in semiconductor manufacturing in photoresist deposition (see discussion of photoresist technology later in this chapter). Moreover, it plays an important role in low-*k* dielectric technology involved in multilayer metallization schemes also considered later in this chapter. Finally, there are several other processes, for instance, in organic semiconductor device manufacture, or in any other application which is using either homogeneous chemical solutions or colloidal solutions precursors to deposit a thin film.

The process of spin coating features certain inherent shortcomings which limit its usefulness in some processes. For instance, it has limited ability to deposit in fully controlled fashion uniform films in the below 50 nm thickness regime, particularly in the case of small or oddly shaped substrates. At the same time spin coating is not practical in the case of large, heavy substrates such as those used in large display manufacture. Furthermore, the spin-on process is not very efficient as only a small part of the material dispensed on the surface as liquid stays there after high rpm spin-off cycle. This is a problem in the case of expensive liquid precursors such as, for instance, colloidal solutions containing nanocrystalline quantum dots.

Considering limitations of spin coating, alternative methods of physical liquid deposition which supplement spin-on technique in the areas where its use is not practical, have been developed over time. Those of potential interest in semiconductor device engineering including mist deposition, micro-spray, and ink-jet printing are introduced as follows.

Mist deposition: A method of coating solid surfaces with liquid precursor with a single-nanometer range thickness control is a method of mist deposition (Price *et al.*, 2009). As name indicates, the liquid in this case is slowly delivered to the substrate in the form of a very fine mist which then uniformly coalesces on its surface. Just like in the case of spin-on process, mist deposition is followed by thermal curing of the film.

The idea behind mist deposition is to convert liquid source material into a very fine mist which is then carried by nitrogen to the deposition chamber where submicron droplets coalesce on the substrate covering its surface with a uniform film of viscous liquid (Fig. 5.17). A liquid precursor is confined in the container from which it is moved into the atomizer by nitrogen pressure. In an atomizer liquid is converted into a very fine

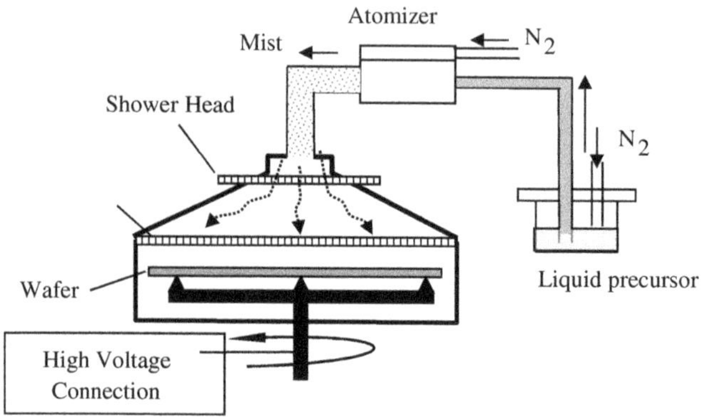

Fig. 5.17. Schematic diagram of the system implementing mist deposition.

mist through interactions with a series of impactors. The average size of the droplet in the mist is about 0.25 μm. but can be smaller for a different impactor configuration. The mist is then carried by nitrogen into the deposition chamber where it coalesces on the surface of a slowly (10 rpm) rotating wafer at room temperature and pressure very close to atmospheric. An electric field can be induced between the grounded field screen and a wafer to enhance mist delivery to the substrate beyond gravitational forces. The film is then subjected to thermal curing at the temperature range 150–300 °C in ambient air which causes evaporation of the solvent leaving on the surface a thin layer of solid (Mumbauer *et al.*, 2004).

What distinguishes mist deposition from other liquid physical deposition techniques is control of the film thickness in the single nanometer regime. At the very slow deposition rate needed for this level of control, this technique is best suited for the deposition of films in the range below 50 nm. Furthermore, mist deposition is independent of the size and shape of the substrate and offers superior to other LPD techniques conformality of coating. With all these features, mist deposition is an effective replacement of the spin-on process in applications for which this last is inherently unsuitable. Due to the very slow rate of deposition, mist deposition process is best suited PLD technique for the bottom-up processes discussed in Section 5.2. This is because mist deposition technique is the one which may respond with required sensitivity to the local changes in surface energy of the substrate established as a result of surface functionalization.

Micro-spray and ink-jet printing: In addition to those discussed above, the thin-film physical liquid deposition methods include commonly used in our daily lives techniques of spraying and ink-jet printing. The difference between spraying and ink-jet printing in daily uses and high-end technical applications is much higher precision of deposition processes in terms of film thickness and line width control in the case of ink-jet printing.

The technique of micro-spray is independent of the size, shape, rigidity, and chemical composition of the substrate. As such, it offers the broadest among LPD techniques range of applications. However, while free from the geometrical restrictions on the substrate of some other LPD techniques, adequate thickness control can be best achieved in the case of micro-spray only for films thicker than about 1 μm. This feature precludes the use of micro-spraying in semiconductor device manufacturing processes where film thicknesses in this range are not used, but opens possibility in some others, including, for instance, manufacturing of large-area solar panels.

In contrast to micro-spraying, which goal is to deposit thin film, the ink-jet printing deposits thin film and establishes its pattern at the same time. As such, it has been investigated as a method allowing direct patterning of the liquid precursor on the surface of the substrate in the micro-scale. The lateral resolution (minimum line width) of the ink-jet printing is in the range of 20 μm to 50 μm, while control of thickness of the patterns ink-jet printed is in the 1 μm range. Despite the geometrical constraints, ink-jet printing is a technique finding applications in the manufacture of the organic semiconductor devices featuring micrometer scale geometries.

Overall, based on the characteristics of various LPD techniques reviewed above, a broadening range of process solutions offered by the liquid physical deposition methods is recognized. It can be assumed that the space for the use of liquid precursors in semiconductor device manufacturing technology will continue to grow in pace with the growing diversity of applications of semiconductor devices.

5.5.6 *Electrochemical Deposition*

The process of electrochemical deposition (ECD) is commonly used across the industries to coat electrically conductive surfaces with thin metal layers. The ECD processes are also known as electrodeposition or electroplating, and in their various embodiments belong to the broader

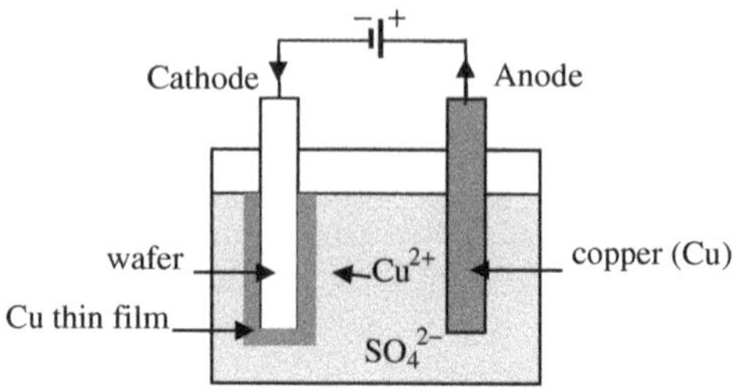

Fig. 5.18. Process of electrodeposition of thin layer of copper.

family of electrophoretic deposition processes. In the case process is run under the constant current conditions, which is usually a case in industrial applications, the term galvanostatic deposition is also used.

The main use of ECD methods in commercial semiconductor manufacturing currently is in the deposition of thin layers of copper acting as interconnect lines in advanced integrated circuits. Process shown in Fig. 5.18 illustrates the workings of this deposition mode using copper anode and electrically conductive silicon wafer (cathode) as a substrate.

In the typical ECD configuration a cathode is a substrate to be coated (silicon wafer) and the anode is a material substrate to be coated with (copper). For the process to work, the electrolyte (electrolytic solution) must be a solution of metal to be deposited which in the case of the process in Fig. 5.18 is a copper sulfide (Cu_2S) mixed with water. Due to the electrochemical reactions taking place at the anode and at the cathode, and with a current flowing between these two electrodes, copper is effectively transferred from the anode to the solution, and then to the cathode covering it with a thin layer of copper.

5.6 Lithography

Lithography is essentially a technique transferring desired pattern to the surface of the wafer. It is a technique of special importance in semiconductor device manufacturing as it defines geometry of semiconductor devices, and thus, determines basic characteristics and performance of each of them. In the case lithographic pattern transfer process is implemented using short-wavelength UV light as a printing energy carrier, the process is known as photolithography (also referred to as optical lithography). As it will be discussed later, various types of photolithography are distinguished based on the wavelength of electromagnetic radiation used (Mack, 2007). In an alternative approach, a focused electron beam is employed as a printing medium, and the technique is known as *e*-beam lithography.

Consider introduction of the *Top-Down* process in Section 5.2.1 as a preamble to the discussion of the lithographic processes in this section. As Fig. 5.1(c–e) indicate, the essence of lithography in semiconductor device manufacturing comes to the definition of desired pattern in the layer of material acting as a pattern transfer material and known as a resist. Depending on the source of energy used for its exposure, it is referred to as photoresist (in the case of UV light used for exposure), or *e*-beam resist (in the case of electron-beam used for exposure).

With a goal being delineation of the desired pattern in the thin-film material deposited on the surface of the substrate (Fig. 5.1(b)), the top-down sequence is completed with the removal of such material in the areas not covered with the resist (Fig. 5.1(f)) using one of the processes considered in Section 5.7.5.

Discussion in this section explains in more details principles the lithographic processes used in semiconductor device manufacturing.

5.6.1 *Implementation of Lithographic Processes*

As stated earlier, the procedure used to transfer the pattern onto the surface of semiconductor wafer (Fig. 5.1) is referred to as *lithography.* The implementation of the lithographic process requires: (i) energy in the form of the as short wavelength as possible UV light (photolithography), or electron beam (*e*-beam lithography) impinging on the processed substrate to initiate pattern defining reactions on its surface, (ii) means to localize the impact of energy on the surface according to predetermined pattern, and (iii) medium responding to this impact so that the effect of the delivered energy can be registered. The issues related to these topics are discussed as follows.

Masked and direct write lithography: Regarding the means used to localize the impact of the energy impinging on the surface of the processed wafer, mentioned earlier techniques of photolithography, and *e*-beam lithography represent two different approaches applied in semiconductor device manufacturing.

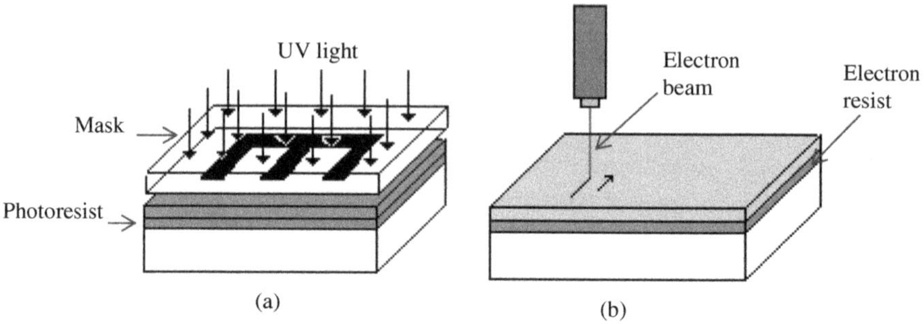

Fig. 5.19. Patterning modes applied in semiconductor device fabrication, (a) mask-based patterning and (b) direct write patterning.

First is using a physical object known as a *mask* which is configured such that it allows UV light to pass through and to expose photoresist only in the areas not covered by the opaque to UV material. This material forms on the surface of the mask a pattern to be transferred (Fig. 5.19(a)). Number of masks used at various stages of the device manufacturing sequence depends on the complexity of the processed device and may vary from just a few in the case of the simple discrete devices to a few dozen in the case of complex integrated circuits.

In the processes where the very high precision of the pattern transfer process is required, and the time needed to delineated the pattern in the layer of the resist is not of the major concern, the desired pattern can be directly written in the resist using highly focused electron beam scanned over the surface of the substrate wafer (Fig. 5.19(b)). This exposure mode is referred to as a direct write lithography and is further considered later in this section in the context of *e*-beam lithography.

Exposure wavelength and resolution of the pattern transfer process: The term *resolution* is concerned with the accuracy with which pattern from the mask is being transferred to the layer of photoresist. When smaller geometrical features need to be created, higher resolution pattern transfer process is required. A limiting factor is the effect of diffraction at the edges of the mask of which adverse effect on the resolution of the pattern transfer process is illustrated in Fig. 5.20. As shown in this figure, the extent of diffraction, and thus resolution of the pattern transfer, depends on the wavelength of light λ passing through the mask and is less pronounced as the wavelength is getting shorter. Consequently, as short at

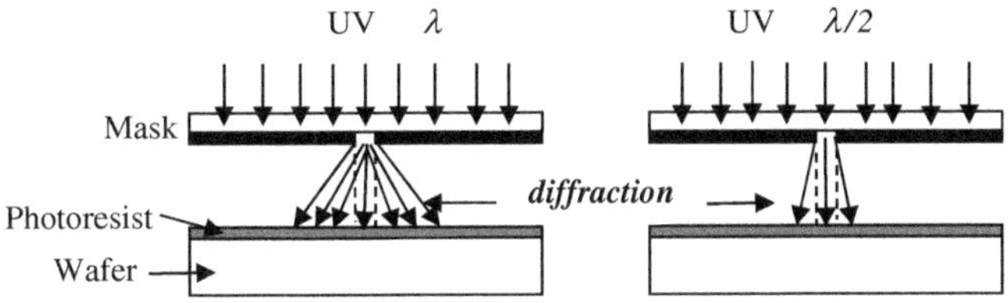

Fig. 5.20. The effect of diffraction defining resolution of the photolithographic process is smaller when the wavelength (λ) of the light used for exposure is shorter (λ/2).

technically viable wavelengths of UV light are used in pattern definition processes in semiconductor device manufacturing in cases where the geometries in the low nanometer range need to be created on the wafer surface.

5.6.2 *Photolithography*

Responding to the need to use short-wavelength, high intensity light in pattern transfer lithographic processes, the UV light representing short-wavelength end of the electromagnetic spectrum is predominantly used as energy carrying medium in semiconductor device fabrication. As a reminder, lithography using UV light for the resist exposure is commonly known as photolithography, or optical lithography.

The UV portion of the electromagnetic spectrum covers the range from about 400 nm down to 10 nm at which point it borders the *X*-ray range. Within the UV spectrum there are selected characteristic lines (wavelengths) identified which feature particularly high intensity and which, depending on the needs driven by the required resolution of the pattern transfer process, are used in photolithography. Based on the illustration the UV spectrum considered earlier (Fig. 4.11), Fig. 5.21 identifies those lines.

In the conventional photolithography, which goal is to delineate geometries in the micrometers range, such as the case of the gate length in the conventional thin-film transistors (TFTs), or contacts width in solar cells, two lines referred to as *g*-line at 436 nm and *i*-line at 365 nm are identified. Variations of the mercury (Hg) arc-based UV lamps are employed to generate desired wavelengths and the photolithographic techniques using them are referred to as *i*-line photolithography and *g*-line photolithography respectively.

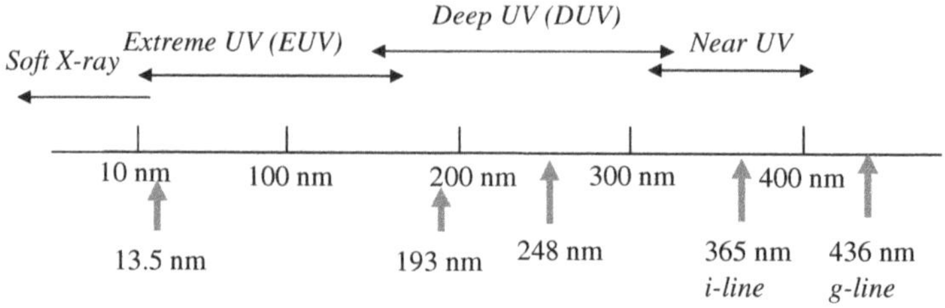

Fig. 5.21. Characteristic lines in UV spectrum which are used in photolithography in semiconductor device manufacturing.

Deep UV (DUV) photolithography: This variation of photolithography is applied when the patterning is concerned with geometrical features in the nanoscale range from 10 nm to 250 nm. Shorter than *i*-line and *g*-line wavelengths in the deep UV (DUV) range need to be used (Fig. 5.21). Generation of such short wavelength UV requires excimer lasers which produce highly uniform, coherent beams of monochromatic light. Among them, other than krypton fluoride (KrF) laser producing 248 nm wavelength, the excimer laser of particular importance in photolithography is argon fluoride (ArF) laser generating high intensity 193 nm wavelength (Fig. 5.21). For all practical purposes this wavelength is a recognized standard bearer in advanced photolithography distinguishing between deep UV and extreme UV photolithography.

Extreme UV (EUV) photolithography: Patterning of geometrical features in nanometer range, requires using UV wavelengths from the extreme UV range within which 13.5 nm wavelength (Fig. 5.21) featuring sufficient intensity can be generated. Technology of EUV photolithography is drastically more complex and costly than DUV photolithography, and thus, is used only in the situations requiring definition of nanometer-scale geometrical features in the manufacture of the leading-edge digital IC.

Generation of the extreme UV wavelengths methods typically involves high-power laser-produced plasma and as such represent entirely different set of challenges than in the case of methods used to produce 193 nm and longer wavelengths. The EUV generation method often exploited involves melting of the tin metal, vapor of which is then excited by high-power CO_2 lasers to form extreme UV emitting plasma. Plasma generating this way is emitting 13.5 nm wavelength intense enough to be used in most advanced pattering technology.

Another feature of EUV photolithography which sets it apart from the other photolithographic methods is concerned with the structure of the masks used. As light in the EUV range is heavily absorbed by all materials, transmissive masks cannot be used in this case. To reduce absorption loses, multi-layer reflective optics in general and multi-layer reflective masks in particular need to be used in the EUV photolithography. To reduce absorption losses further, parts including EUV sources and patterning systems must be housed in high-vacuum chambers. All these measures make a reflective mask very structurally complex material system, more detailed considerations of which are beyond the scope of this discussion in this *Guide*. Consequently, in the follow-up subsection focus is on transmissive masks used in DUV and UV photolithography.

The lithography technique using "soft" X-rays instead of UV and known as X-ray lithography (Fig. 5.21) is not actively pursuit because by using wavelengths very close to the extreme UV range, it does not offer meaningfully better resolution of the pattern transfer process than EUV lithography.

Photomask: The masks used in UV and DUV photolithography are called photomasks and are transmissive (Fig. 5.22). Depending on the way photomask is used in the photolithographic exposure systems, term reticle is used in reference to the photomasks applied in the projection printing considered later in this section.

A photomask consists of two parts (Fig. 5.22). The first one is a transparent to any given UV wavelength used and is commonly referred to as a blank. To assure adequate transparence to UV light, the blank, which is at the same time acting as a mechanical support for the opaque parts of the mask, is manufactured using highest quality quartz.

The part of the mask that needs to block off UV light, in other words opaque part of the mask, is in the form of the thin film of optically very

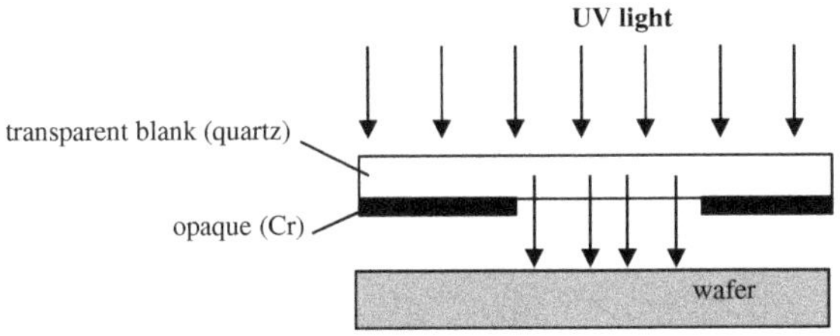

Fig. 5.22. Working of the transmissive photomask.

dense material such as most commonly, chrome (Cr). Besides optical density, opaque material needs to be structurally homogeneous to assure as limited as possible atomic-scale roughness of the edges of the opaque lines (line edge roughness, LER). In the case of applications less demanding in terms of process resolution, photomasks made on glass instead of quartz as a transparent part, and photoemulsion instead of chrome as an opaque part can be used.

If needed to increase resolution of the pattern transfer process, the photomask as shown in Fig. 5.22 can be modified by adding to the blank part of the mask a feature that can change the phase of the light passing through by 180°. The resulting phase shift mask (PSM) is a standard in advanced processing. The phase-shifting is typically accomplished by adding a thin layer of the properly selected material (for instance, silicon nitride, Si_3N_4) featuring precisely determined thickness to the transparent parts of the mask. Alternatively, phase shifting of the light passing through the mask can be accomplished by locally decreasing thickness of the blank to the precisely determined depth.

As a reminder of the point made earlier, an alternative to the transmissive masks considered here, are the reflective masks where the UV light is reflected from the mask and the pattern is projected on the surface of the wafer covered with photoresist. The choice between transmissive and reflective masks is determined by the wavelength of the UV light used with the latter applied in the case of the extremely short wavelength UV exploited in the extreme UV photolithography.

Advanced photomasks described above are fabricated using direct-write *e*-beam lithography discussed later in this section.

Photoresist: The material deposited on the surface of the wafer for the purpose of registering the local impact of the energy carried by UV light is called a resist. The resists used in photolithography are known as photoresists (Figs. 5.19(a) and 5.20).

In terms of chemical composition, the resists are organic compounds formulated such that they respond with highest sensitivity to the specific wavelength of UV light. It means that differently formulated photoresists are used in conventional photolithography, and DUV lithography, for instance. Still, different resists are used in *e*-beam lithography discussed later.

Photoresist is a material which is sensitive to the UV light, or in other words, the UV light is promoting in the layer of photoresist photochemical reactions which alter its solubility in the developer. The class of photoresists in which UV irradiation changes material from initially insoluble in the developer to soluble are referred to as positive photoresists. On the other hand, the class of photoresists in which UV irradiation changes material from initially soluble in the developer to insoluble are referred to as negative photoresists. The positive photoresist allows higher resolution of the pattern transfer process, and thus, is used to delineate very small patterns. On the other hand, the negative photoresist is more sensitive to UV light which means that it requires shorted exposure time, and as such allows higher throughput of the photolithographic processes.

In starting form, photoresist is a highly viscous polymer-based liquid which upon deposition on the surface of the wafer, most commonly by the spin-on process considered in Section 5.5.5, is solidified by low temperature curing causing evaporation of the solvent.

Typically, deposition of photoresist is preceded by the deposition using the same spin-on process of the thin layer of material assuring adequate adhesion of photoresist to the substrate (adhesion promoter). In addition, thin film of UV absorbing material called a bottom antireflective coating (BARC) is often spin-coated on the surface of the wafer prior to photoresist deposition to prevent undesired reflection of the UV light passing through photoresist from the surface of the wafer.

In addition to their function as a photosensitive materials, photoresists also play a role of the materials masking etching during subsequent to photolithography subtractive processes (see Section 5.7). This means that photoresists need to be very chemically resistant materials maintaining their characteristics even when exposed to aggressive chemistries.

At the conclusion of any pattern definition sequence, layer of photoresist must be removed from the surface of the wafer. The resist stripping process is among the most often applied subtractive processes in the fabrication of semiconductor devices. As such, it is discussed further in Section 5.7.5.

5.6.3 *Exposure Techniques in Photolithography*

In the conventional photolithography, step implemented prior to photoresist exposure, regardless of the exposure technique used, the process during which mask is aligned with the pattern already existing of the surface of the wafer. A mask alignment process, carried out automatically with

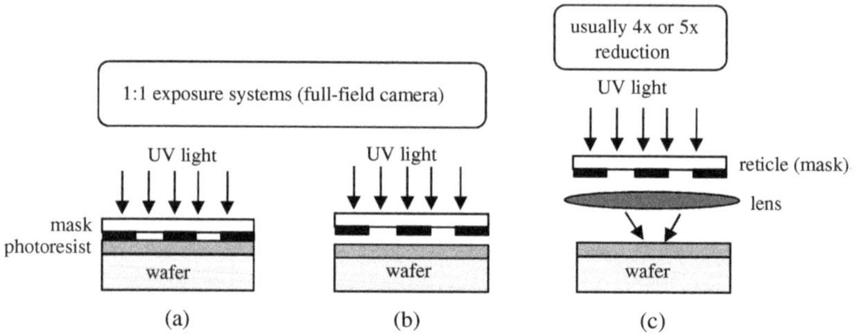

Fig. 5.23. (a) Contact printing, (b) proximity printing, and (c) projection printing.

assistance of the laser beams, uses alignment marks on the mask to position it according to the patterns created during earlier patterning steps. Once mask is aligned with the pattern on the wafer, exposure of photoresist is taking place

There are various ways exposure of photoresist using transmissive masks can be implemented. Three among them, namely contact printing, proximity printing, and projection printing are schematically presented in Fig. 5.23.

Contact printing: The process involves the full-field exposure of the wafer. During contact printing, mask remains in physical contact with the surface of the wafer covered with photoresist (Fig. 5.23(a)), thus, reducing the adverse effect of the diffraction of the light passing through the mask illustrated in Fig. 5.20. In this way, patterns as small as about 0.5 μm can be delineated with contact printing. The downside of this approach is a potential damage to the mask and the wafer resulting from the mechanical interactions likely to occur when these two materials are aligned with respect to each other, and then brought to the intimate contact. Also, relatively low throughput of the contact printing defines its usefulness in mass production of semiconductor devices.

Proximity printing: Similarly, to contact printing, proximity printing is an exposure technique in which the entire wafer is exposed to the UV light (full-field camera) as shown in Fig. 5.23(b). In contrast to contact printing, however, the mask in this case is positioned very close to the wafer (separation typically on the order of 20 μm) but is not in physical contact with the surface of the wafer. In this way possible damage of the mask resulting from the contact with photoresist on the surface of the wafer is prevented, but at the expense of the resolution of the pattern transfer process. Due to the gap between the mask and the wafer, pattern distorting diffraction effect illustrated in Fig. 5.20 does not allow delineation of the patterns smaller than about 2 μm.

Projection printing: Projection printing is an exposure technique commonly used in photolithography applied in the industrial manufacturing processes of semiconductor devices. It represents a departure from the principles of the contact and proximity printing because pattern on the mask is projected on the surface of the wafer not directly, but through the complex lens system positioned between the mask and the wafer in the proximity of the latter (Fig. 5.23(c)). With this configuration, the image created by UV light passing through the mask can be manipulated toward improved accuracy (resolution) of the pattern transfer using adequately designed lens system positioned within a short distance from the wafer surface.

As opposed to full-field camera used in contact and proximity printing, projection printing uses step-and-repeat camera commonly referred to as stepper which reduces pattern on the mask and prints it on the surface of the wafer. Then, wafer is stepped to a new location and exposure is repeated. The sequence is repeated as many times as needed to reproduce the image on the mask on the surface of the entire wafer.

Immersion lithography: This technique is designed to improve resolution of the pattern transfer process at the set wavelengths of the UV light used for exposure. For instance, at the shortest UV wavelength of 193 nm used in DUV lithography, patterns as small as 10 nm can be resolved using proper resolution enhancement techniques. One such technique involves phase-shift masks considered earlier.

Among resolution enhancing techniques used in sub-22 nm photolithography, immersion lithography is the well-established solution. Immersion photolithography involves projection printing in which space between final projection lens and the wafer in the stepper (Fig. 5.23(c)) is filled with water. Water features refractive index $n = 1.44$ while the refractive index of air is $n = 1$. By using medium through which UV light is passing featuring increased refractive index, numerical aperture (NA) of the optical system increases, thus, allowing smaller critical dimensions (CD), or a minimum feature size, to be delineated on the surface of the wafer.

Multiple printing lithography: Yet another approach to resolution enhancement is concerned with the way patterns are printed in the layer of photoresist. In the case of the very small, high-density, complex patterns to be created, a single UV exposure may not be enough to accomplish high-resolution of pattern transfer process. A solution is a multiple printing lithography during which more than one printing step assures uniform exposure of all, even the smallest patterns.

It is quite common in advanced semiconductor device manufacturing that more than one resolution enhancement technique is being used to accomplish desired result in terms of the minimum feature size created on the surface of the wafer. For instance, multiple patterning can be used in combination with immersion lithography to get desired results.

5.6.4 *Electron-Beam Lithography*

Among non-optical lithographies, or lithography techniques using energy carrying exposure medium other than short-wavelength UV light, electron-beam lithography (*e*-beam lithography) is a preferred solution due to its advantageous characteristics including direct-write capability.

The *e*-beam lithography is using a finely focused beam of electrons to write a pattern directly into the resist without using mask (Fig. 5.19(b)). An electron beam down to the single nanometers in diameter can be formed allowing finer patterns to be formed than with conventional DUV photolithography. A resist in this case is known as an electron-beam resist (*e*-beam resist) and is formulated differently than photoresist used in photolithography to respond specifically to the kinetic energy carried by electrons impinging upon it rather than to the energy of short-wavelengths UV light as it is in the case of photolithography.

The reason why patterns same as the diameter of the *e*-beam cannot be exposed using this lithography mode is the discussed earlier proximity effect. As a reminder, proximity effect is caused by the scattering of electrons in the beam penetrating the resist as well as emission of secondary electrons from the solid underneath the resist (see Section 4.5.3). As a result, area of resist subjected to interactions with energized electrons is larger than the diameter of the incident electron beam.

Despite the limitations imposed by the proximity effect and relatively low throughput, *e*-beam lithography is the method of choice in mask fabrication processes, in the manufacture of specialized devices, as well as in prototyping, process development, and research.

5.7 Subtractive Processes

The lithographic processes discussed in the previous section, engrave the desired geometry in the layer of resist. In this way the first stage of the pattern definition procedure is completed. During the second stage, the material not covered by the resist will be removed during the subtractive process referred to as etching (Fig. 5.1(e)). The need for the high-performance etching processes cannot be overestimated. This is because challenges that had to be resolved to ensure high resolution of the lithographic processes discussed above would serve no useful purpose if they weren't followed by the equally high precision etching processes.

In this section, the most important aspects of etching technologies in semiconductor device fabrication are considered. First, general characteristics of etch processes are reviewed and various etching modes are presented. Then, methods of liquid phase, (wet), and gas phase, (dry), etching will be discussed. Finally, techniques used to remove layer of resist following etching know as resist stripping will be briefly reviewed.

5.7.1 *Characteristics of Etch Processes*

The etching processes in semiconductor engineering are commonly considered in terms of the etch selectivity/non-selectivity, and isotropy/anisotropy. Selective is an etch which is featured by the high etch rate of one material only, while its interactions with other materials on the surface and in the near-surface region of the wafer are negligible. The result of the selective etch is shown in Fig. 5.24(a) where material *A* as etched

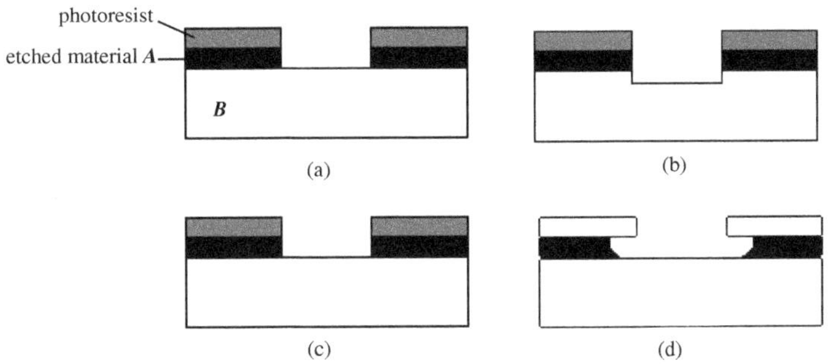

Fig. 5.24. Characteristics of etch processes (a) selective, (b) non-selective, (c) anisotropic, and (d) isotropic.

off without any etching of underlying material B and no erosion of the resist film. In contrast, the etch rate in the case of nonselective etching is relatively independent of the chemical composition of the material exposed to etching agent. Figure 5.24(b) illustrates non-selective etch where after removing material A etching process continued by etching material *B*.

Besides interactions of materials with etching agent to which they are exposed, an issue of interest is directionality of etching process. An anisotropic etching features high directionality of material removal as indicated in Fig. 5.24(c) by straight sidewalls of the window etched in material *A*. In contrast, isotropic etching occurs at the same rate in any direction. As a result, besides etching of material *A* in the direction normal to the surface, equally effective lateral etching is taking place (Fig. 5.24(d)). Resulting undercutting is responsible for the distortion of the pattern in the oxide as compared to the pattern in the resist.

While the non-selective etching processes are in general not desired, and thus, efforts to make etching as selective as possible are commonly attempted, the choices between isotropic and anisotropic etching modes are based upon the requirements of any given process. The control between those is done by selecting appropriate etching methods as discussed later in this section.

Various etching methods are also categorized based on the nature of interactions responsible for etching. Etching process is considered purely chemical (chemical etching) when removal of the material is accomplished solely through the chemical reactions resulting in the formation of compounds that are soluble in the etching agent in the case of liquid-phase etching (wet etching) or are volatile in the case of the gas-phase etching (dry etching). An intermediate in terms of the phase of the etching medium is a chemical vapor-phase etching in which etchants are delivered in the gas phase, but etching reactions are taking place in the liquid-phase.

In Table 5.1 different etching modes are related to the etch characteristics. As seen, chemical etching can be implemented using liquid-, vapor-, and gas-phase chemistries. In each case etching can be made selective, but at the same time is highly isotropic as there is no reason for the chemical etching reaction to proceed unidirectionally.

Purely physical etching on the other hand, accomplishes material removal through the transfer of momentum between energized species

Table 5.1. Etching modes applied in semiconductor manufacturing.

	CHEMICAL	CHEMICAL/PHYSICAL	PHYSICAL
WET *Liquid-phase*	selective isotropic	—	—
DRY/WET Vapor-phase	selective isotropic	—	—
DRY *Gas-phase*	selective isotropic	selective anisotropic	

bombarding the surface, for instance, argon ions, and atoms of the bombarded material. In this case, atoms or clusters of atoms of the etched material are ejected in the same way as in the case of sputtering process discussed in Chapter 4 and earlier in this chapter.

As far as physical etching is concerned, its implementation in the liquid-phase is not possible because no species in the liquid can be accelerated sufficiently to sputter exposed solid.

In the high-density, small geometry device manufacturing, such as advanced logic integrated circuits, etching is highly localized, and methods are required that will be sufficiently anisotropic and selective at the same time. Hence, both physical and chemical interactions must be simultaneously taking place during etching (Physical/Chemical etching in Table 5.1). Due to the physical interactions involved, such etching mode can only be implemented by means of gas-phase chemistries. On the other hand, as will be shown later in this section, there are applications in which etch anisotropy is not required. In such cases, etching can be driven only by chemical reactions which can be implemented using either wet, vapor, or dry ambient depending of the needs of any given etching process.

An important parameter defining subtractive processes, regardless of how they are implemented, is the etch rate, or in other words the rate at which removal of the material by etching is progressing. For any specific etch chemistry/material combination the etch rate is different and needs to be precisely determined before actual implementation of the etch process in device manufacturing. At a set etch conditions, the etch rate is strongly dependent on the chemical composition and structural features of the etched material, and thus, based on the etch rate measured, changes in the basic characteristics of the etched material can be detected.

5.7.2 *Wet Etching*

Wet etching operations are based on the general principles of wet processes in semiconductor manufacturing which were discussed in Section 4.3. Also, considerations of wet cleaning processes earlier in this chapter shed additional light on the way wet etching operations involved in semiconductor device manufacturing are carried out.

Special role in wet etching processes is played by the water rinse which is needed not only to remove products of the etch reactions from the surface of the wafer, but it is also used to stop etch reaction at the desired moment. Furthermore, as the discussion in Chapter 4 indicates, any wet process requires a thorough rinsing and drying of the processed wafers and the wet etching is no exception.

The compositions of the wet etching chemistries are very diverse, specific to the materials to be etched, and to the purpose of a given etch process. A number of wet etch chemistries can be identified for each material whether it is a semiconductor, dielectric, or a conductor. Some were in general terms identified earlier, but an extensive listing of even only the most important material-specific wet etch chemistries would go beyond the scope of this *Guide*. So, just as an example, selected few etch chemistries used to process common in semiconductor technology materials are considered below.

When wet etching of silicon is concerned, various water solutions of chemistries involving halogen acids, for instance, hydrochloric acid (HCl), are selected depending on the etching mechanism of choice. Often, a process based on the oxidation-reduction cycle involving water solution of nitric acid (HNO_3) and HF is used to etch silicon. The water solution of HF is also broadly used chemistry in silicon dioxide, SiO_2, etching making hydrofluoric acid HF one of the common liquid-phase chemicals used in semiconductor manufacturing.

Wet etching of compound semiconductors is more challenging and, in some cases, silicon carbide SiC, for instance, is ineffective. As a result, etching of such materials relies on the gaseous etch chemistries. In the case of III–V compounds such as GaN, GaAs or InAs, often processed in various combinations as ternary compounds, etching using wet chemistries involves broad range of chemicals of even a superficial overview of which is beyond the framework of this discussion.

In contrast to readily etchable SiO_2, some other dielectrics used in semiconductor device manufacture require relatively aggressive chemistries to be etched. A prime example here is silicon nitride (Si_3N_4) which requires phosphoric acid, (H_3PO_4) at 180 °C to be etched at the reasonable rate. Also, effectiveness of wet etching of metals used for contacts and interconnects in semiconductor devices varies from material to material. For instance, etching of aluminum (Al) can be relatively easily implemented using water solution of phosphoric acid (H_3PO_4), acetic acid (CH_3COOH), and nitric acid (HNO_3). On the other hand, wet etching of metals such as copper or refractory metals creates technical problems which are often worked around by using alternative solution such as, for instance, damascene process discussed in Section 5.9.3.

Due to its basic characteristics, wet etching is particularly suitable for the implementation of what is known as a preferential etching processes. In the course of preferential etching the etching process occurs at a significantly higher rate along certain crystallographic planes in the crystalline solid. The process is most used to identify structural defects in the single-crystal semiconductor materials used in device fabrication.

In terms of implementation, wet etching operations in mass production of semiconductor devices, most commonly involve immersion technology using wet benches discussed in Chapter 4.

5.7.3 *Vapor-Phase Etching*

Vapor-phase etching is identified here as a separate class of etching modes because of its distinct characteristics and applications. It combines isotropy of wet chemistry with penetrability of the tight geometrical features possible with the reactants in the gas phase. In the case of vapor-phase processes the reactants are delivered to the etched surfaces in the form of the mixture of vaporized liquid chemical and the vapor of water, or an organic solvent. Vapor is condensing on the surface where it gets involved in the etch reaction in the liquid-phase. Since the substrate is maintained at the temperature above room temperature, typically at 40–70 °C, products of the etch reaction are removed from the surface in the vapor-phase. The temperature is used to control etch rate and to prevent formation of the solid residues on the etched surfaces (Torek *et al.*, 1995).

The prime example of the vapor etching is the process using vapor of the anhydrous HF (AHF) and the methanol, or ethanol, mixed in the vapor-phase. In the vapor-phase, AHF:alcoholic solvent mixture can penetrate tight geometrical features which due to the surface tension cannot be penetrated by the liquid HF solutions.

The AHF:alcoholic solvent process is used most often in the MEMS release processes (Fig. 3.26) in which etching of the sacrificial oxide in the intricate lateral geometries needs to be isotropic and selective. It is also used in the removal of the ultra-thin spontaneously grown native oxide from the silicon surface prior to contact deposition process.

Due to its unique characteristics, compatible with the needs of the oxide etching in the narrow, lateral channels, the AHF:alcoholic solvent etch is an integral part of the silicon MEMS device technology (Erdamar *et al.*, 2008).

5.7.4 *Dry Etching*

Dry etching involves gases which are free of any water vapor or vapor of gases which are reacting to produce water. In the majority of practical applications dry etching processes involve electrically active species generated through the electric discharge as discussed in Section 4.5.2. This is the reason why dry etching methods involve plasma as a process driving medium and density of plasma plays a role in defining efficiency of the etch process. Depending on the etching mode, electrically active species can be chemically neutral, for instance, argon ions Ar^+ or chemically reactive such as, for instance, chlorine ion Cl^-. Unlike wet and vapor etching, dry etching can be implemented in either isotropic chemical etch, or anisotropic and non-selective physical etch. When needed, dry etching can be configured to combine chemical and physical etch interactions.

As Fig. 5.25 shows dry etching mode is controlled by the pressure of reactant gases and energy of etching species which is increasing toward purely physical etching. In the roughly 1–100 Torr pressure range and with

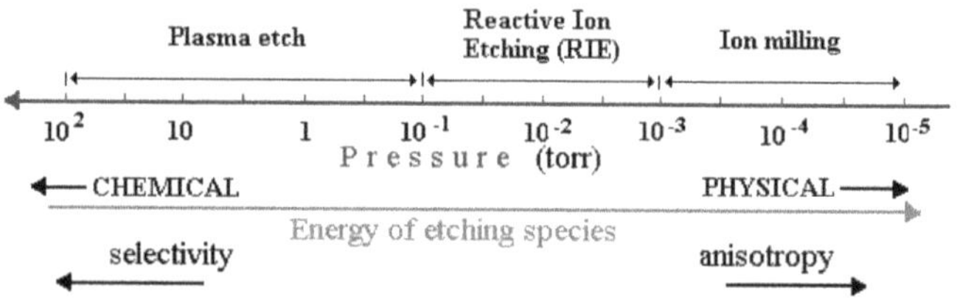

Fig. 5.25. Pressure and energy of etching species used in various dry etching modes.

etching species featuring low energy (no acceleration toward etched material) etching is strictly chemical and is referred to as plasma etching.

On the other hand, at the pressure in the range of 10^{-3}–10^{-5} Torr and with etching species carrying kinetic energy (acceleration toward etched material), etching becomes purely physical, and thus, anisotropic and non-selective. As such it is known as ion milling, and also as sputter etch.

The dry etching method combining characteristics of plasma etch and ion milling which is implemented at the moderate energy and pressure (10^{-3}–10^{-1} Torr) is a method of Reactive Ion Etching, (RIE). The RIE is the most often employed dry etching method in semiconductor device manufacturing. General characteristics and applications of various dry etching modes are briefly summarized as follows.

Plasma etching: It involves an isotropic subtractive process in the course of which etching occurs through the chemical reactions between plasma generated etching species and etched material. In the absence of the forces promoting motion of the etching species toward the wafer surface (Staffa *et al.*, 1995), plasma etching depends on the random interactions between the etchant and the wafer. In this way plasma etching is analogous to the conventional wet etching process (Fig. 5.26(a)). Because etching species arrive at the etch surface carrying very little kinetic energy, the plasma etching process is not causing any pronounced surface damage.

Plasma etching is commonly used in photoresist stripping operations (see Section 5.7.5) and in other processes where anisotropy of etching is not required while damage of the etched surface needs to be avoided. In terms of reactors configuration, plasma etches are implemented using

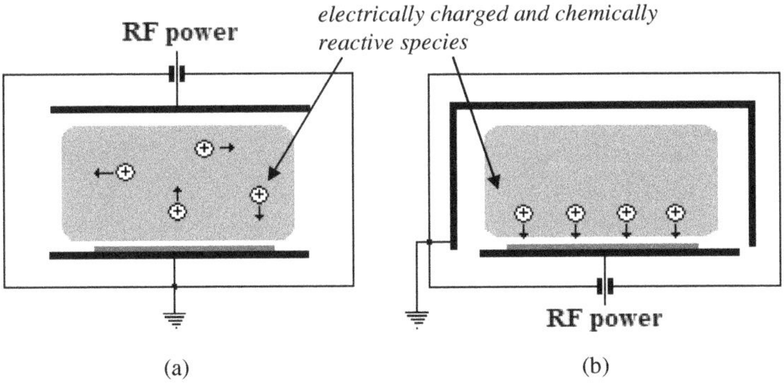

Fig. 5.26. (a) Plasma etch reactor and (b) Reactive Ion Etch, RIE reactor.

either among parallel plate reactors considered in Chapter 4, or barrel reactors configured to carry out plasma-driven batch processes, or remote plasma reactors (Staffa *et al.*, 1995).

Reactive ion etching (RIE): This etching mode provides the best combination of anisotropy and selectivity, and thus, is the most applied etching technique in pattern defining applications. Due to the physical component involved in the etch process, RIE usually leaves etched surfaces physically damaged and often contaminated with the species penetrating the near-surface region of the etched material.

RIE differs from the pure plasma etching in terms of the configuration of the etch reactor and the way RF power is applied (Fig. 5.26). Together, it alters distribution of potential in the system such that the plasma generated chemically reactive ions are accelerated toward the wafer. This changes the nature of interactions between the etching species and the etched material from purely chemical in the case of plasma etch (Fig. 5.26(a)), to the combination of chemical and physical in the case of RIE (Fig. 5.26(b)).

As the most versatile etching process, the RIE can be used to pattern any material involved in semiconductor device manufacturing regardless of their chemical composition and crystallographic structure. This includes some refractory metals and copper, for instance, which are difficult to etch using other techniques. The diversity of materials that can be processed by means of RIE brings about a broad range of possibilities in terms of the selection of etch chemistries. Composition of the RIE chemistries needs to be determined depending on the chemical makeup of the material to be etched, as well as specific goals of the etching process in device fabrication sequence. In some applications ultra-slow, multicharged ions can be used for etching (Le Roux *et al.*, 2004).

In the implementation of RIE processes commonly used are parallel-plate reactors which are typically equipped with plasma density increasing features such as those provided by the Inductively Coupled Plasma (ICP) technique (Fig. 4.9(a)).

Magnetically enhanced reactive ion etching (MERIE): As indicated earlier, application of the magnetic field confines electrons in the plasma and results in increased ionization efficiency, and thus, increased plasma density which leads to the increase of the etch rate. Depending on how electromagnetic field is coupled with plasma, various types of MERIE reactors, including Inductively Coupled Plasma (ICP), ECR plasma, and helicon plasma reactors are distinguished.

Overall, enhancement of the RIE processes using magnetic field is a broadly adopted solution in the dry etching technology.

Deep reactive ion etching (DRIE): Deep RIE is a variation of RIE processes geared specifically toward prolonged etches creating μm-range lateral and horizontal features involved in the engineering of MEMS devices. Commonly used in these applications is a Bosh process which was developed specifically for the purpose of deep etches required in MEMS fabrication. The process is carried out in a series of steps which are accompanied by the formation of the layer of polymer on the exposed sidewalls of the etched features. The purpose of the protective polymer layer is to prevent the undesired lateral etching during lengthy DRIE runs. Upon completion of the DRIE process a layer of polymer is removed using plasma etching.

Ion milling: In contrast to plasma etch and RIE, ion milling is a purely physical dry etching mode indicated in Fig. 5.25. The ion milling is basically a process of a directional sputtering of the exposed material during which its etching occurs solely through the physical interactions with chemically inactive ions such as Ar^+ accelerated toward the etched material. Naturally, the process is highly anisotropic and non-selective as sputtering coefficients of various materials tend to be similar. The term ion "milling", schematically illustrated in Fig. 4.10, adequately reflects the nature of the process.

Atomic layer etching (ALE): It is the gas-phase etching mode which escapes classification of dry etching techniques illustrated in Fig. 5.25. It needs to be listed here, however, due to its importance in advanced semiconductor processing. Just like discussed earlier atomic layer deposition (ALD) which allows controlled conformal deposition of a few nm thick films, ALE allows removal of the material with the same atomic-scale precision. However, layer-by-atomic layer removal of the material is technically more challenging than equally precise deposition. This is because in contrast to isotropic, conformal ALD process, which is mostly independent of the chemical composition of the substrate as it creates its own surface chemistry, atomic layer etching needs to be anisotropic and highly selective while being dependent on the chemical composition of the etched material. Similarly to ALD, to accomplish atomic layer precision of the material removal process, the ALE reaction needs to be self-limiting.

In short, Atomic Layer Etching is a specialized process and its more detailed considerations are beyond the scope of the introductory discussion in this *Guide*.

5.7.5 *Resist Stripping*

As it was established earlier, in the typical top-down patterning process resist (photoresist in the case of photolithography) is used as a mask during etching which means that after etching is completed, the resist must be entirely removed from the surface of the wafer. The process known as resist stripping is a step which concludes pattern definition sequence as shown in Fig. 5.1. As an organic based material, photoresist is removed by dissolution in liquid solvent such as acetone, or in the gas phase by oxidation in the oxygen plasma. In this last case photoresist stripping is often referred to as a photoresist ashing process. In the batch processes of semiconductor wafers, the barrel reactors are commonly used to carry out resist stripping operations.

The resist striping is a demanding process as it is expected to leave the processed surface free of any solid residues and contaminants. The challenge comes when the photoresist is used as a mask during ion implantation (see Section 5.8.2). The species implanted in photoresist during ion implantation are altering its chemical composition to the point where its exposure to oxidizing chemistries is not enough to assure its removal without leaving residues on the processed surface. If encountered, additional cleaning steps are required to cope with this challenge.

5.8 Selective Doping

There are several ways by which controlled addition of small amounts of alien elements may drastically affect basic properties of the host semiconductor materials. For instance, gallium arsenide (GaAs), will assume ferromagnetic properties when manganese (Mn) atoms are introduced into its structure and change it into GaMnAs.

A special role among the processes designed to alter properties of a solid by introducing into its structure alien elements is played by the process referred to as doping used in fabrication of semiconductor devices. (Ghandi, 1983). In the follow-up discussion the term doping is being used specifically in reference to the process introducing alien elements referred to as dopants (sometimes term impurities is used instead) into a given semiconductor material, silicon, for instance, to alter its electrical conductivity, and/or to change its conductivity type (see Fig. 1.1(b) and discussion in Section 1.2).

The ability to control electrical conductivity of semiconductor materials in general, and conductivity type (*n*- or *p*-type) in particular, is a foundation of any semiconductor device manufacturing technology. As discussion in Chapters 1 and 2 has shown, electrical conductivity of semiconductor material can be established either by homogeneously adding dopants during material growth or selectively (locally) into carefully defined by means of patterning parts of the substrate wafer during device manufacturing (Fig. 5.5(b)). The single-crystal growth processes, as well as epitaxial deposition of semiconductors, or CVD of doped polycrystalline or amorphous semiconductors are examples of the former.

The discussion of the doping processes in this section is focused on selective doping of the host semiconductor material during device fabrication. The term selective doping is understood here as a process which forms laterally and vertically confined regions either *n*- or *p*-type within uniformly pre-doped semiconductor substrate.

Two methods used to carry out processes of selective doping in semiconductor manufacturing are based on the processes of diffusion and ion implantation.

5.8.1 *Doping by Diffusion*

Diffusion is a process which by its nature can only be initiated in the presence of concentration gradient of dopant atoms. Another precondition for the diffusion to occur is elevated temperature at which host material needs to be maintained. Only then can atoms in its lattice be substituted by the diffusing dopant atom (substitutional diffusion), form a bond with the host atoms, and then play the role of either acceptor or donor. As pointed out earlier in this volume, in the case of silicon, element from group III of the periodic table, boron (B) specifically, is serving as an acceptor, or *p*-type dopant, while elements from the group V, phosphorus (P), arsenic (As), and antimony (Sb) can be selected, depending on the process, to act as *n*-type dopants.

Alternatively, some elements can diffuse in the host semiconductor, such as silicon, by moving in between host atoms and without any bonding with the host atoms (interstitial diffusion). For obvious reasons fast diffusing elements of this type cannot be used as dopants of silicon. Metals such as copper and gold are examples of fast diffusants in silicon.

By nature, the process of doping by diffusion involves high-temperature operations, and hence, its use is limited to the thermally stable elemental semiconductors and is of no use in the case of some thermally unstable compound semiconductors.

Consecutive steps involved in the formation of the doped region by diffusion, for instance, addition of the *p*-type dopant to *n*-type semiconductor material to form a *p–n* junction, are schematically illustrated in Fig. 5.27 using *n*-type silicon substrate featuring donor concentration in the bulk N_B as an example, and boron as a *p*-type dopant.

The first set of operations aims at the formation of the masking layer of silicon dioxide (SiO_2) patterned using photolithographic process (Fig. 5.27(a)). It defines the region which dopant atoms will penetrate (diffuse in) to form a *p–n* junction. Following formation of the masking

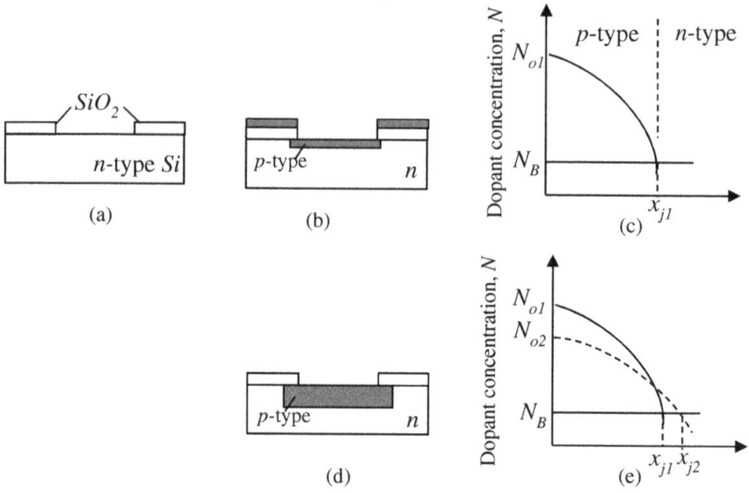

Fig. 5.27. Illustration of the doping by diffusion in silicon (a) a layer of SiO_2 acting as a mask is formed and patterned on the Si surface, (b) *p*-type region is formed by diffusion, (c) corresponding dopant distribution, (d) *p*-type region after drive-in diffusion and doped oxide removal, (e) change in the dopant distribution following drive-in diffusion.

oxide the wafer is exposed at high temperature to the boron (dopant) and oxygen containing ambient to form a thin layer of oxide containing high concentration of boron. The process is referred to as pre-deposition and is also known as an unlimited source diffusion. As the pre-deposition step is carried out at the elevated temperature, boron, due to the concentration gradient, penetrates the near surface region of the silicon substrate forming a *p–n* junction at the junction depth x_{j1} (Fig. 5.27(b)). The resulting distribution of boron is shown in Fig. 5.27(c) which depicts a junction depth x_{j1} at the point where concentration of diffused-in acceptors equals concentration of donors in the *n*-type substrate, i.e. $N_A = N_B$. It also shows the surface concentration N_{o1} which is determined by the solid solubility limit of boron in silicon at the temperature of pre-deposition process.

The removal of the dopant-rich oxide from the wafer surface (Fig. 5.27(d)) prevents addition of new dopants into the substrate, but does not prevent redistribution of the dopants already introduced into silicon in the case when the thermal budget of the additional thermal process exceeds thermal budget of the pre-deposition step. Such dopant redistribution process known as a drive-in, or limited source diffusion, lowers the surface concentration from N_{o1} to N_{o2} are drives deeper the *p–n* junction as shown in Fig. 5.27(e). The drive-in process can be applied deliberately for the purpose of modification of dopants distribution, or can be an undesired byproduct of the thermal treatment to which wafer is subjected for reasons not related to the process of doping by diffusion. Either way, once dopant atoms are introduced into semiconductor substrate, then they are subject to the same redistribution processes regardless of the doping technique used.

The process of doping by diffusion is by definition an elevated temperature operation and commonly involves conventional resistance heating furnaces (Fig. 4.6) which are often referred to as diffusion furnaces. In the processes requiring shorting of the time of the pre-deposition diffusion, so that the shallow *p–n* junction can be formed, the Rapid Thermal Processing (RTP) tools allowing low thermal budget process are used (Fig. 4.7).

As the shape of the doped regions in Fig. 5.27 indicates, diffusion is an isotropic process which means that the lateral diffusion expands the doped region underneath the edges of the masking oxide (Fig. 5.27(d)). While such a pattern distortion will have a negligible impact on the device configuration in the case of devices featuring relaxed geometry, it cannot be tolerated in the case of devices and circuits featuring very tight geometries. For that reason, in the manufacture of integrated circuits with *nm*-scale dimensions, alternative to diffusion selective doping technique of ion implantation is used.

5.8.2 *Doping by Ion Implantation*

This doping mode is implemented using technique of ion implantation introduced in Section 4.5.3 and illustrated in Fig. 4.10(c). In this *Guide* ion implantation is discussed as an alternative to diffusion method used to introduce dopants into semiconductor material for the purpose of altering its electrical conductivity and/or changing its conductivity type. Due to its inherent characteristics allowing superior to diffusion control over the distribution of implanted dopant atoms, ion implantation has become a method of choice in any semiconductor doping processes requiring precise control over the lateral and vertical dopants distribution.

In the process of ion implantation, ions of the desired dopant (for instance, boron which is a *p*-type dopant of silicon) are generated in the plasma from which they are then extracted and accelerated toward the substrate to be doped. The ions impinge of the surface of the substrate upon which desired pattern is defined by the masking material (for instance, silicon dioxide (SiO_2) in Fig. 5.28(a)), penetrate its near surface region losing energy through inelastic collisions with the atoms in the lattice of the implanted material, and eventually come to the rest at the certain distance from the surface. Due to the random nature of inelastic collisions involved, not all implanted ions come to the rest at the same distance from the surface, and thus, doped region spreads over the certain distance from the surface.

Figure 5.28(b) shows the profile of the implanted dopants in the parts not protected by the masking oxide, or any other sufficiently thick film of masking material, photoresist, for instance. Distribution of the implanted dopants features concentration peak N_p which is located at the distance from the surface denoted by *Rp* and called a projected range. The projected range is a function of implantation energy while concentration of dopants and concentration peak N_p in the implanted region are determined by the time of implantation and the density of ion beam current expressed by number of ions in the beam crossing per second unit area of the implanted surface. Time and beam current define parameter of implantation known as an implantation dose. The key feature of the implanted dopant profiles shown in Fig. 5.28 (a) and (b) is that they reproduce

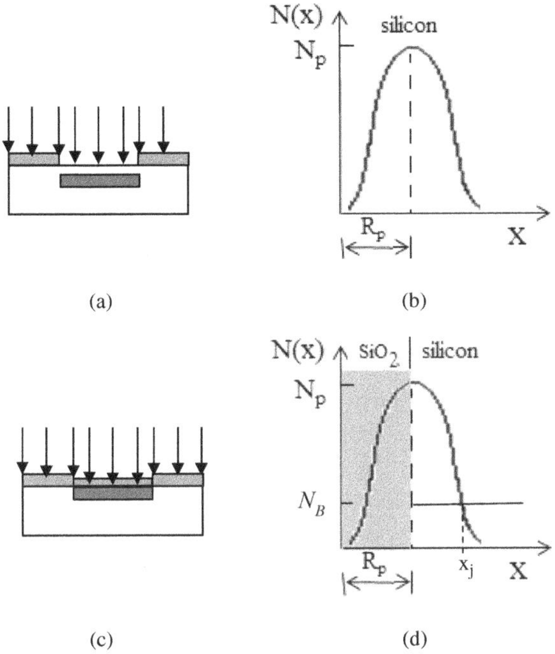

Fig. 5.28. (a) Implantation into parts of the wafer not covered by the masking oxide, (b) resulting distribution of implanted dopants in silicon, (c) implantation through the thin oxide on the surface, and (d) resulting dopants distribution.

pattern created in the masking material with negligible lateral distortion making ion implantation a selective doping technique of choice in any doping processes involving nanometer-scale geometries.

To eliminate incompletely doped layer between the surface and the implanted region (Fig. 5.28 (a) and (b)), implantation can be carried out through the film of an oxide formed on the surface Fig. 5.28(c)), the thickness of which corresponds roughly to the projected range R_p (Fig. 5.28(d)). In this way, only a part of the implanted ions reaches silicon substrate and with proper adjustment of the oxide thickness and implantation energy, concentration peak can be positioned exactly at the surface and the *p–n* junction formed at the desired depth x_j (Fig. 5.28(d)). Following implantation, oxide implanted with dopant ions is etched off the surface of the wafer.

A deleterious effect accompanying ion implantation is damage of the crystal structure sustained by the region immediately adjacent to the surface through which high energy ions are passing before coming to the rest. To remove this damage and to restore crystallographic order, as well as to activate implanted dopants, process of ion implantation is followed by the low thermal budget post-implantation anneal carried out by means of the rapid thermal processing (RTP) tools. RTP tools are used in this case to avoid dopant redistribution that would otherwise occur, much like in the case of drive-in process discussed earlier in relation to the selective doping by diffusion process (Fig. 5.27(e)).

Also associated with doping by ion implantation is the effect of channeling which has an adverse effect on the distribution of implanted ions. It occurs when implanted ions hit open "channels" in between the atoms in the implanted single-crystal lattice. By not being subjected to collisions with atoms in their lattice sites, channeling ions come to the complete rest significantly deeper than other ions. To minimize the effect of channeling it is a common procedure to position implanted wafer at a certain angle with respect to the direction of the ion beam.

The simplified representation of the building blocks comprising a typical ion implanter is shown in Fig. 5.29. Dopant ions generated in the plasma chamber (1) are extracted from the plasma and passing through a mass analyzer and separator (2) which removes from the beam ions other than the dopant atoms. Then, the beam of ions featuring uniform energy distribution across its cross-section is formed, and ions are accelerated in the acceleration tube (3) to reach desired energy which defines projected range, and thus, implantation depth. Section (4) of the implanter tool includes instrumentation allowing beam scanning and blanking before it finally impinges on the wafer in the wafer process chamber (5) maintained at the adequately reduced pressure.

There are various classes of ion implanters designed to meet various manufacturing goals. For instance, for deep implants (see, for instance, a

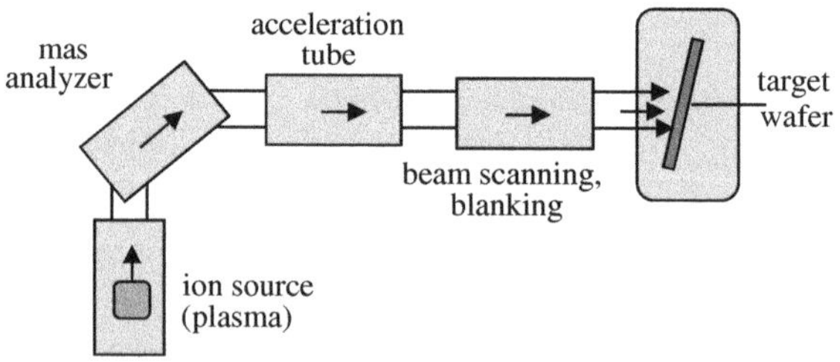

Fig. 5.29. Schematic illustration of the building blocks of an ion implanter.

SIMOX process used to fabricate SOI wafers considered in Chapter 2, or trench isolation implantation), high-energy implanters allowing implantation energy above 1 MeV need to be used. Low-energy, high-current implanters that could operate at the energy below 1 keV, but need to feature high beam current, are used to form shallow junctions with high dopant concentration at the surface (for instance, source and drain regions in the MOSFETs).

To summarize an overview of ion implantation technology in semiconductor manufacturing, the following points need to be made. First, ion implantation, in contrast to diffusion, is essentially a room temperature process carried out in vacuum which forms sharp, well controlled doping profiles. Second, a low-thermal budget post-implantation anneal removing implantation damage and activating dopants is an integral part of the selective doping by ion implantation procedure. Third, once implanted into the semiconductor material, dopant atoms are subject to the exact same laws of the diffusion in solids (dopants redistribution) as those introduced by the process of diffusion in the course of pre-deposition runs. Finally, the ion implantation is a materials engineering method used in various applications beyond semiconductor doping processes, for instance, in metal surface processing operations.

5.9 Processing of Contacts and Interconnects

In semiconductor terminology terms *contacts* and *interconnects* are used in reference to electrically conductive materials needed to flow electric current in and out of the device and to provide electrical connection between devices in the circuit (Fig. 5.30) As low as possible resistivity and thermal stability are key requirements concerning properties of

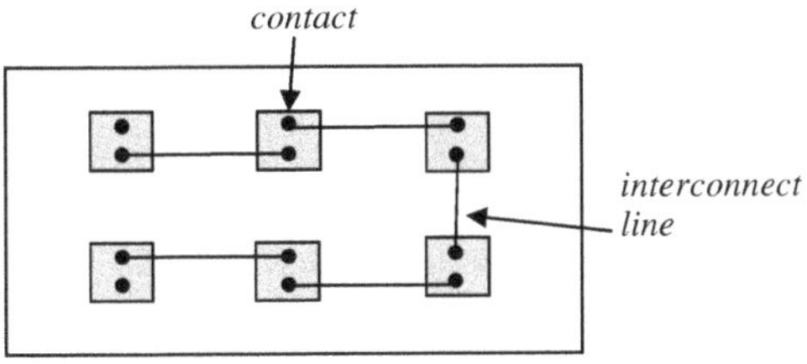

Fig. 5.30. Illustration of contact and interconnects in semiconductor devices integrated on the chip into circuit.

materials used to process contacts and interconnects in semiconductor device manufacturing.

As indicated earlier in this *Guide*, semiconductor manufacturing procedures, particularly in the case of the manufacture of advanced devices such as complex integrated circuits, are divided into parts referred to as Front-End-Of-the Line, or FEOL processes, and Back-End-Of-the Line, or BEOL processes. An operation forming the first metal contact on the wafer surface is considered to be a step completing the FEOL part of the process, while at the same time being the first step in its BEOL portion of the process. Processing of interconnect lines in the integrated circuits falls squarely into BEOL part of the manufacturing sequence and is subject to different requirements comparing to contacts formation procedures during the FEOL part of the process.

Discussion in this section is concerned with the fabrication technology aspects of contacts and interconnects processing in semiconductor devices and circuits, including deposition and patterning procedures. As a reminder, contacts and interconnects related topics were considered in Section 2.10 of this volume. Furthermore, selected fundamental issues related to the properties of ohmic contacts, which play important role in the overall device manufacturing scheme, as well as Schottky contacts, were discussed in Section 3.3.

5.9.1 *Contacts*

As should be apparent from the discussion in this *Guide*, no semiconductor device can function based solely on the current, or voltage controlling interactions within semiconductor material. Every semiconductor device, with no exception, needs to be equipped with ohmic contacts assuring undisturbed flow of current in and out of the device (Fig. 5.30). As low as possible resistivity of the contact material is required to assure negligible series resistance introduced into to the current flow path. Selected measures taken to minimize contact resistance considered earlier were concerned with the formation of silicides (Fig. 2.26), and processing of implanted contacts where charge transport between heavily doped semiconductor and metal is controlled by tunneling. In the former case the process of self-aligned silicide, "salicide" in short, which is commonly used to form source, gate, and drain contacts in the MOSFETs without designated patterning step, is a prime example of the use in silicide technology in silicon device manufacturing.

In addition to the selection of the low resistivity contact material featuring desired work function, the condition of the semiconductor surface prior to metal deposition plays a role in assuring ohmic contact characteristics. This is because in addition to the resistance of the contact material, a factor potentially increasing contact resistance, is a spontaneously formed film of residues involving native oxides and organic contaminants that are commonly present on the substrate surface prior to contact deposition. Regardless of whether contact is acting as an ohmic contact, or Schottky contact, or whether it is transparent to light or non-transparent, intimated physical contact between metal and semiconductor is desired (Fig. 5.31(a)). Presence of the ultra-thin (typically not exceeding 1 nm) residue film at the interface between contact material and semiconductor (Fig. 5.31(b)) results in a major distortion of the contact characteristics manifesting itself in the increasing contacts resistance and its overall poor performance.

To minimize the impact of the residue film, adequate processing of the surface upon which contact if deposited is an integral part of the contact formation process. Most commonly, it is concerned with an operation aimed at the removal of the ultra-thin oxide spontaneously grown on the semiconductor surface during the wafer handling. In the case of silicon, it involves either brief immersion in the weak HF(1):water(100) solution, or exposure to the vapor of an anhydrous HF mixed with alcoholic solvent such as ethanol (see Section 5.7.3). Alternatively, the process may include a brief sputter etch of the surface oxide in vacuum system immediately prior to metal deposition.

Somewhat different requirements regarding contact material are formulated depending on device's operational characteristics. For instance, temperature resistant metals are needed to make contacts in power devices

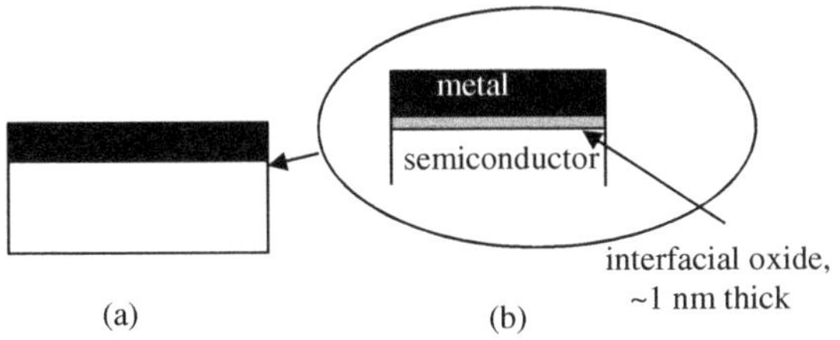

Fig. 5.31. (a) Intimate physical contact between metal and semiconductor is (b) disturbed by the interfacial residue film.

where high current may increase temperature of the contact material significantly. In the case of devices interacting with light in turn, contact materials that are transparent to light are required.

Thin-film contacts formation processes, including deposition and etching, are briefly considered below for the contacts non-transparent and transparent to light. As pointed out earlier, in either case contact's series resistance needs to be as low as possible to assure desired device performance.

Non-transparent contacts: Contacts non-transparent to light are primarily metals, for instance, aluminum (Al) and gold (Au), or metal alloys such as discussed earlier titanium nitride, TiN, or silicides discussed earlier. The choice regarding the deposition technique used is typically between PVD method of sputtering and the one of the CVD methods among which Atomic Layer Deposition (ALD) introduced in Section 5.5.4 is used in the case of particularly demanding metal deposition steps. The thermal evaporation method can only be applied in the case of low-melting point metals and is relatively infrequently used in mass production of semiconductor devices. Instead, sputter deposition is a method of choice among PVD techniques. The same considerations regarding deposition method apply also to metal deposition step in the process of silicide formation (Fig. 2.21).

Special needs are imposed by the deposition procedures of metals in the case of which either PVD or CVD techniques are not producing high-quality films. In the case of copper, for instance, which is commonly used as an interconnect metal (see discussion later in this section), a method of electrochemical deposition (Fig. 5.18), preceded by deposition of seed layer by means of sputter deposition is a commonly adopted solution.

Deposition of the thin-film metal is just a first step in the process of contact metallization. The other one is pattern defining etching process which in the case of some metals can be a challenge. The most common metal in semiconductor technology which is aluminum (Al), can be easily etched in the liquid-phase using mixture of phosphoric acid (H_3PO_4), nitric acid (HNO_3), and water. On the other hand, gold (Au) can only be etched using *aqua regia* which is a 3:1 mixture of hydrochloric (HCl) and nitric acids (HNO_3) which because of its extreme chemical reactivity is incompatible with semiconductor process infrastructure. For that reason, gold contacts (for instance, to some III–V compound semiconductors) are patterned using lift-off procedure discussed in Section 5.2.3.

Etching of copper, either dry or wet, assuring adequate control of the process and producing sharp nanoscale patterns is difficult. Therefore, in the case of interconnects in which copper finds its main use, definition of the copper lines geometry is based on the method in which etching of copper is not required (see damascene process in Section 5.9.3). Regarding other metals of interest, some refractory metals such tantalum (Ta) and molybdenum (Mo) cannot be etched using wet chemistries. Instead, plasma etch or RIE using gaseous carbon tetrafluoride (CF_4) is being used. Tungsten (W) can be etched using the same gas phase chemistry, but in addition it can be removed in the liquid-phase using mixture of nitric (HNO_3) and hydrofluoric (HF) acids. The same wet chemistry is effective in etching titanium nitride (TiN).

Transparent contacts: As the name indicates, transparent contacts need to be electrically conductive and optically transparent. They are often used as back contacts in light emitting devices. Typically, they belong to the broad family of Transparent Conductive Oxides (TCO).

The best-established representative of electrically conductive materials which are transparent to light is indium tin oxide (ITO). The ITO is deposited predominantly by means of a PVD method of sputtering using high-purity ITO targets. As far as etching of ITO is concerned, a solution of hydrochloric acid (HCl) and water with small amount of nitric acid (HNO_3) added is an etch chemistry of choice in this case.

5.9.2 *Interconnects*

As the name indicates, the purpose of interconnects is to connect individual devices such as transistors processed into the semiconductor chip into an electronic circuit (Fig. 5.30). Interconnect lines are in the form of properly patterned thin-film electrical conductors most commonly metals. Alternatively, in selected applications superconductors can be used as interconnect lines. In yet another variation of the interconnect scheme, signal distribution around the circuit can be accomplished using light and waveguides instead of electrical current and metal interconnect lines. In the follow-up discussion metal interconnect lines are considered as the representatives of the mainstream interconnect technology.

In the case of the circuits featuring relaxed geometries, and thus, remaining operational with a single-level interconnect scheme, aluminum is typically used as the interconnect metal. Aluminum interconnects can be readily processed using the same deposition and etching methods as those used to form contacts considered above.

Situation is different in more complex, high-density ICs where shortcomings of aluminum such as, for instance, electromigration are coming to play and instead, copper is broadly used as an interconnect metal. Furthermore, as an earlier discussion in this *Guide* has revealed, due to the limitations imposed on the scaling of the interconnect lines in high density ICs, multilevel metallization scheme needs to be implemented.

Multilevel metallization: Reasons for which interconnect network in high density integrated circuits is realized using multilevel metallization scheme (Fig. 3.23), instead of single-level interconnect system, were explained in Section 3.6. Here, selected aspects of the methodology employed to process such multilevel structure are outlined.

To facilitate explanation of the key concepts pertaining to the multilevel metallization technology, Fig. 5.32 in the simplified fashion identifies key elements of the multilevel metallization system used in advanced ICs. Its three elements include metal lines, interlevel dielectrics (ILD), and vias, also referred to as plugs. While for the sake of simplicity Fig. 5.32 shows two metal levels only, in actual high-density ICs, number of metal levels is commonly above ten. What it means is that the multilevel metallization scheme represents a complex material system which includes materials featuring distinctly different properties, and which is expected to carry out its function at the increased temperature, and with a high-density currents flowing across the system.

Processing of multilevel metallization systems is at the very core of the BEOL part of the IC manufacturing sequence and features stringent requirements regarding performance of materials and processes.

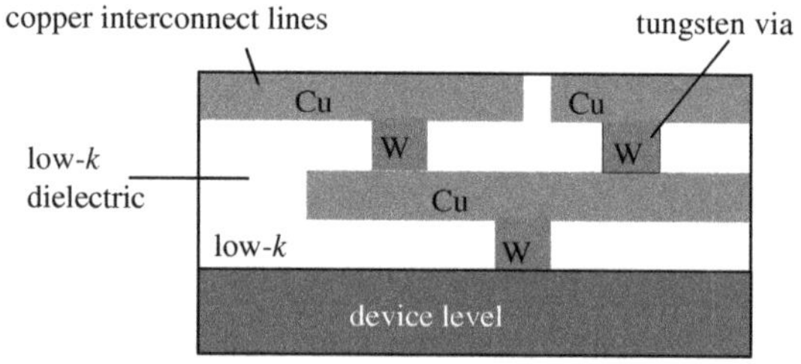

Fig. 5.32. Multi level/multi material IC interconnect system.

Considering the former, the key element is concerned with the selection of metal used to form interconnect lines. As mentioned earlier in this *Guide* (see Section 2.11) the metal of choice in this application is copper (Cu). It features very high electrical conductivity and what is important in interconnect technology, it is not affected by the effect of electromigration plaguing some other highly conductive metals, most notably aluminum.

Regarding processes used in the copper deposition, key consideration is the temperature at which film is formed. Deposition temperature in this case needs to be as low as possible to prevent any alteration of the intricate multilevel metallization scheme. The low-temperature technique commonly employed to deposit copper is electrochemical deposition illustrated in Fig. 5.18 which, when used in conjunction with dielectrics in multilevel metallization scheme (Fig. 5.32), needs to be preceded by the deposition of the thin-film seed copper layer by means of sputtering. Alternatively, a Metalorganic Chemical Vapor Deposition (MOCVD) of copper involving complex metalorganic precursors is used, although, it is employed primarily in the process of Through-Silicon Vias (TSV) discussed in Section 5.9. The reason for MOCVD use in this case is that it allows deposition of copper at the temperatures as low as 200–300 °C.

Another metal used in the processing of multilevel metallization is tungsten (W) commonly employed to form vias (plugs) depicted in Fig. 5.32. Deposition of tungsten in this application is implemented by means of Chemical Vapor Deposition with tungsten fluoride (WF_6) used as a gaseous source of tungsten. With the low-pressure CVD, proper selection of reactants, and possible plasma enhancement, temperature of tungsten deposition can be lowered to below 500 °C.

Interlevel Dielectrics (ILD): A separate set of challenges in interconnect technology is presented by the selection of materials and processing of low-*k* dielectrics acting as interlevel dielectrics (Fig. 5.32). As indicated earlier (Section 2.10.3), incorporation of *k*-lowering porosity to the material acting as ILD is a common practice in high-end ICs manufacturing. As far as deposition of ILDs is concerned, almost uniquely it is carried out by means of low-temperature CVD in the case of gaseous precursors, or by liquid physical deposition (LPD), primarily using spin-on process, in the case of precursors which are in the form of high viscosity liquids. Regardless of the deposition method used, inclusion of porosity into the film for the purpose of lowering its dielectric constant k is an important objective.

5.9.3 *Damascene Process*

The specific requirements and needs of the multilevel interconnect processing can be accomplished when patterning, deposition, and chemical-mechanical planarization (CMP) steps are integrated into what is known as a damascene process. In this process the interconnect metal lines are delineated in dielectrics isolating them from each other not by means of photolithography and etching, but with the use of the CMP principles of which are explained in Fig. 4.12. As Fig. 5.33 shows, in the damascene process first the interconnect pattern is photolithographically defined in the layer of low-k dielectric (Fig. 5.33(a)), then metal is deposited to fill resulting trenches(Fig. 5.33(b)). In the last step, excess metal is removed by means of CMP (Fig. 5.33(c)).

A modified version of the conventional damascene process shown in Fig. 5.33, the one which is typically used in IC manufacturing, is a dual-damascene process. In dual damascene process two interlevel dielectric patterning steps and one CMP step create a pattern which would require two patterning steps and two metal CMP steps when using conventional damascene process.

As pointed out in Section 5.4, process known as a scrub cleaning is an inherent part of any chemical-mechanical planarization/polishing procedures such as those involved in damascene process, as well as in wafer thinning operations considered in the following section. Due to the heavy contamination of the wafer with the polishing slurry remaining on the surface, post-CMP cleaning operations following to involve mechanical

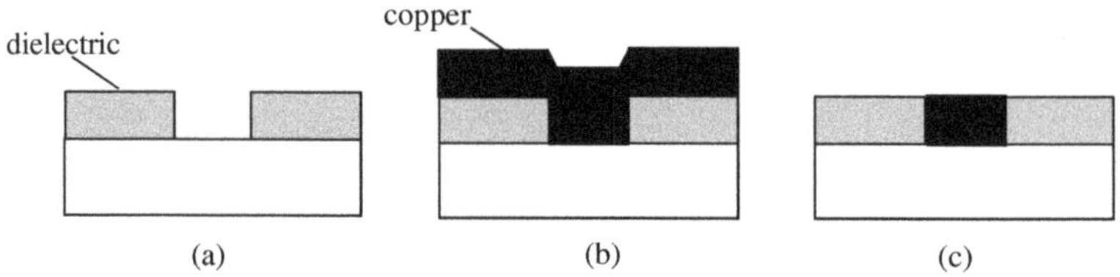

Fig. 5.33. Simplified illustration of the damascene process (a) patterned dielectric, (b) copper deposition, and (c) excess copper removal by polishing.

interactions using soft brushes, or aggressive megasonic agitation to stimulate chemical cleaning reactions.

5.10 Assembly and Packaging

The process of assembly and packaging involves the very last set of operations in the manufacture of any electronic and photonic semiconductor device. In the follow-up discussion, operations converting completely processed semiconductor chips fabricated on semiconductor wafer into self-contained, packaged devices that can be connected with larger electronic or photonic systems are briefly considered.

The goal of assembly and packaging process is to convert a chip being a part of semiconductor wafer (Fig. 5.34(a)) coming out of the fabrication process into packaged functional device which can be installed in electronic or photonic circuitry (Fig. 5.34(b)).

There is no standard assembly and packaging procedure that can be implemented with each semiconductor device regardless of whether it is an integrated circuit, discrete transistor, solar cell panel, light-emitting diode, or MEMS device, for instance. As each type of semiconductor device requires specific to its function assembly and packaging procedures, a review of all possible schemes would go beyond the scope of this *Guide*. Therefore, a brief overview in this section is focused, as an example, on selected issues involved in the assembly and packaging procedures employed in the mainstream integrated circuits manufacturing. considered to be an adequate representation.

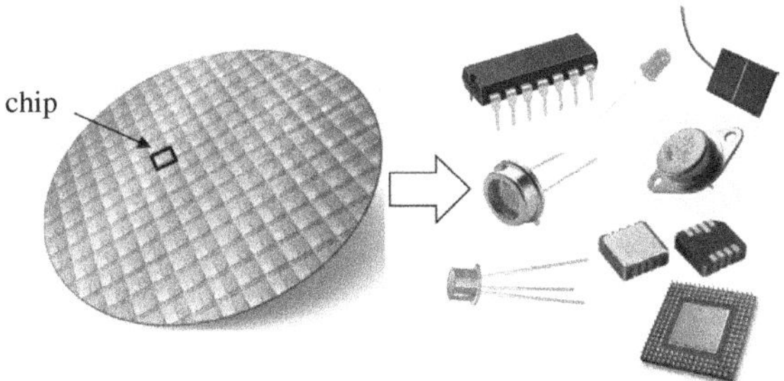

(a) (b)

Fig. 5.34. From the processed wafer containing hundreds of fully processed individual chips (a) to the packaged devices (b).

5.10.1 *Assembly and Packaging Process*

Figure 5.35 attempts to summarize and to illustrate in the simplified fashion key steps involved in the assembly and packaging of an integrated circuit starting with a processed wafer containing in its structure fully functional chips (Fig. 5.35(a)). The very first step in the sequence is concerned with electrical measurements of the functional and parametric characteristics of every chip on the wafer (Fig. 5.35(b)). As a result of such fully automated tests, chips which do not meet expected specifications are identified and marked. The results of testing determine a number referred to as a manufacturing yield which represents a percentage of the chips performing according to the specifications. Low manufacturing yield is an indication of the inadequate performance of the fabrication process calling for the appropriate remedial steps to be taken.

The next step involves operation referred to as dicing which by means of cutting using fine diamond blades or scribing, depending on the thickness of the wafer, separates wafer into individual chips (Fig. 5.35(c)). The chips which were determined earlier as malfunctioning are discarded. Each correctly working chip is subsequently mounted in the package which in the case shown in Fig. 5.35(d) is a Pin Grid Array (PGA) type of the package commonly used in the manufacture of

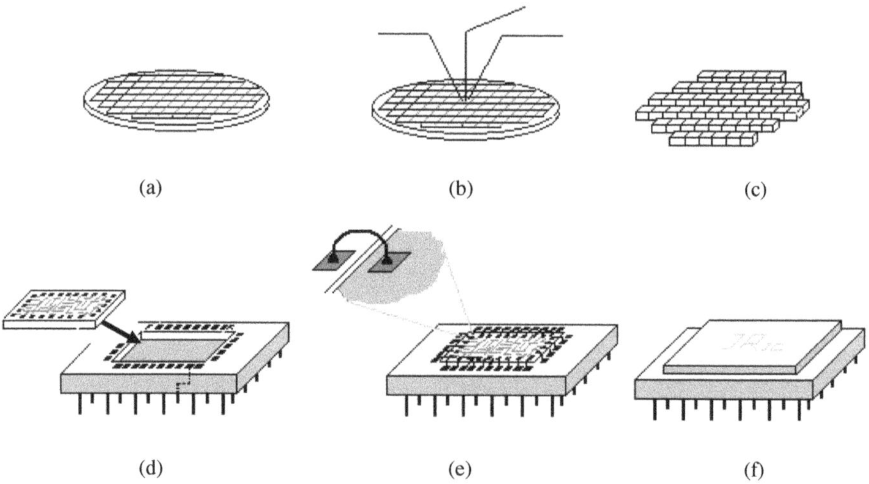

Fig. 5.35. Assembly and packaging of the IC where (a) fully processed wafer is (b) subjected the electrical testing, then (c) diced (separated into individual chips0, (d) installed in the package, (e) wire bonded, and (f) encapsulated in the sealed package.

integrated circuits. Once in place, chip is electrically connected to the package using wire bonding technique (Fig. 5.35(e)) The last step in the sequence encapsulates and permanently seals the chip protecting package (Fig. 5.35(f)).

5.10.2 *Evolution of Semiconductor Packaging Technology*

The package is an indispensable link between semiconductor devices and outside electronic and photonic systems. No progress in device technology, nanotechnology or not, will ever have any impact unless performance of the package will keep on improving in sync with improving device performance. Therefore, semiconductor device packaging technology is continuously evolving to accommodate emerging needs across all types of devices discussed in Chapter 3 of this *Guide*.

A good example of the evolution of the IC packaging technology schematically illustrated in Fig. 5.36 are solutions implemented in response to the continuously increasing ICs functionality attained by packing more transistors per chip. Increased packing density cannot be accomplished solely by scaling transistor geometry, but needs to be supported by the increased size of the chip, and thus, footprint of the package. Neither is desired as the former increases length of the interconnect lines causing increase of the interconnect delay, while the latter increases footprint of the package prohibiting incorporation of the complex electronic systems into relatively low-volume gadgets such as, for instance, smartphones.

A solution to these challenges is found in 3D chip stacking inside the package which involves interconnection of multiple chips by means of through silicon vias (TSV) technology (Fig. 5.36). The TSV technology requires thinning of the wafer prior to dicing to some 70 μm (as compared

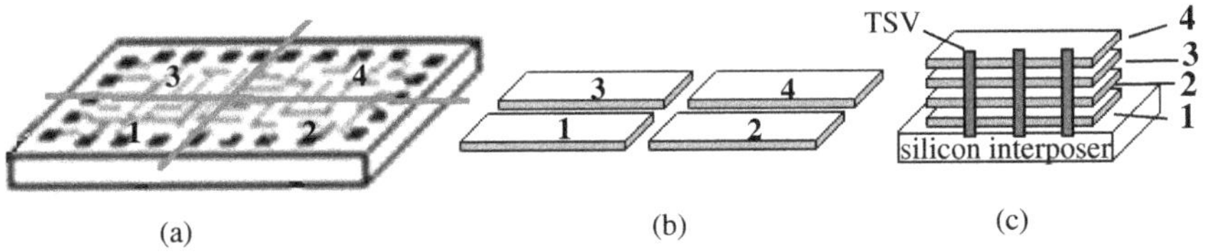

Fig. 5.36. Schematic illustration of the concept of 3D packaging where (a) processed chip is (b) thinned and separated into functional blocks which are then (c) stacked and interconnected using TSV technology.

to a typical large wafer being upwards of 700 μm thick) by means of CMP, and then separating of the large chip into functional blocks such central processing unit (CPU), memory, and analog/mix signal chips (Fig. 5.36(b)). In the "via first" approach the chips are then etched using DRIE process (see Section 5.7.4) to create vias. Upon alignment and bonding of the chips stacked on the silicon interposer acting as an electrical interface with the package (Fig. 5.36(c)), vias will be filled with copper typically by means of the MOCVD process. The result is a stack which at the reduced thickness of the chips can be thinner than the conventional wafer, and thus, can be accommodated by the conventional PGA package.

There are several advantages of the 3D chip stacking including reduced interconnect capacitance and the delay time (due to the shortened interconnect length) resulting in the increased processing speed of the entire circuit, as well as lower power consumption, and reduced noise. It also accommodates the needs of the complex System-on-Chip (SOC) layout without increasing the footprint of the package. What is not improved is a heat management capability of the 3D stack which makes 3D packaging favoring speed of operation rather than power handling capabilities of the circuit. Finally, a major advantage of 3D integration is that it allows heterogeneous integration which means that, for instance, silicon chips and gallium arsenide chips can be integrated in the same package.

In addition to cutting edge IC technology, the needs of advanced MEMS, LEDs, power devices, and silicon photonics all call for innovative solutions regarding packaging technology. In the case of MEMS devices, a trend toward standardized lower cost, smaller packages continues. In the process, TSV-based technology is challenging more traditional wire bonding approach.

Yet another set of challenges concerns the packaging of solar cells. The case of solar cells is special because unlike other types of semiconductor devices, solar cells by definition are permanently subjected to interactions with elements. This includes exposure to sunlight, temperature changes between day and night, which can be significant depending on the location and season, as well as rain and moisture, all of which are having a major effect of the lifetime of the cell. In particular, part of the package known as front sheet is especially important because it is expected to remain transparent to sunlight and non-reflective regardless of the environment. All in all, the cost of packaging in solar cells manufacturing encompasses larger part of the total manufacturing cost than in any other semiconductor device.

What also needs to be pointed out are the special demands regarding package design and materials used in the case of power devices. This is because in the case of silicon carbide (SiC) Schottky diode, for instance, temperature higher than 400 °C can be encountered. Special needs in terms of heat dissipation and cooling capabilities of the package are a major concern in this case.

In conclusion, importance of the packages in semiconductor device technology needs to be underscored. While not directly involving semiconductor materials, packages provide a link to the outside world without which outstanding characteristics of semiconductor materials could not be used in any practical application.

5.10.3 *Heterogeneous Integration*

As Fig. 3.21 shows progress in semiconductor engineering technology is often expressed in terms of the number of transistors per mm^2 of the chip area. Progress in this regard is accomplished by continued reduction of transistor's geometry, changes of transistor architecture, and use of most advanced technologies.

Different approach to improved functionality and lower manufacturing cost of semiconductor integrated circuitry at is pursued by heterogeneous integration (Ravi *et al.*, 2026). Instead of traditional approach in which chips originating from the same wafer substrate are mounted into the same package (Fig. 5.36), heterogeneous approach integrates into the same package chips fabricated using different materials (e.g. silicon and gallium nitride), and designed to perform various functions, for instance, digital, analog, and electro-mechanical. Furthermore, each may come from the different fabrication facilities, originating from substrate wafers different in terms of shape and chemical composition shape, and manufactured using different design rules and geometry (Fig. 5.37).

As shown in Fig. 5.37, heterogeneous integration implemented in planar configuration (3D heterogeneous integration is a viable option) fits into a package chips performing different functions and possibly chips manufactured in the different fabrication facilities. Due to heterogeneous integration electronic circuitry featuring improved design flexibility and superior performance as compared to the circuits based on printed board technology can be manufactured at the reduced cost.

In the light of the promise it shows, various aspects of heterogeneous integration are drawing significant attention and are vigorously pursued

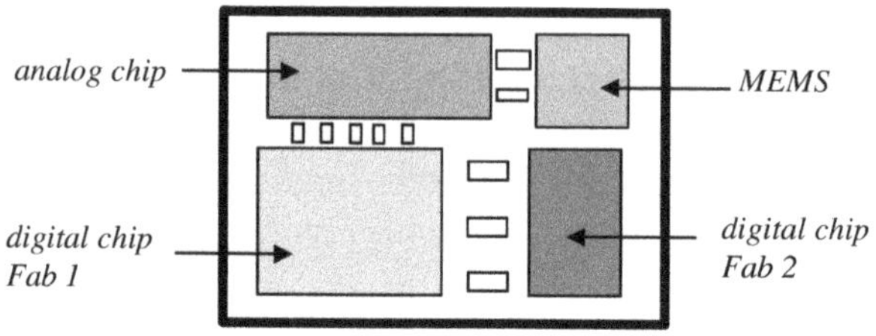

Fig. 5.37. Illustration of heterogeneous integration in planar configuration in which chips of different origin are mounted and enclosed in the same package.

(Swaminathan, 2025). Overall, it can be assumed that heterogeneous integration will contribute significantly to the further growth of semiconductor engineering.

Review Questions

5.1. Graphically present sequence of processing steps included in typical top-down process.
5.2. How mechanical mask is being used to define specific pattern on the surface of the wafer?
5.3. What are the main contaminants in semiconductor device manufacturing environment? Explain the difference between surface cleaning and surface conditioning.
5.4. Consider kinetics of the process of thermal oxidation of silicon distinguishing between its two different stages.
5.5. What are the general rules defining CVD process?
5.6. Consider difference between PVD processes based on evaporation and sputtering.
5.7. Show the sequence of processing steps included in photolithography.
5.8. What exposure techniques are used in photolithography?
5.9. Explain the difference between UV and electron beam lithography.
5.10. For what purposes etching processes are applied in manufacturing of semiconductor devices?
5.11. What is the difference between RIE and plasma etching?
5.12. Explain the difference between doping by diffusion and by ion implantation.

5.13. Using schematic diagram explain the role of contacts and interconnects in integrated circuits technology.
5.14. Present steps included in typical assembly and packaging from finished wafer to packaged chip ready for inclusion in an electronic circuitry.
5.15. Explain what is heterogeneous integration.

Chapter 6

Semiconductor Materials Characterization and Process Monitoring

6.1 Introduction

There is a strong correlation between condition of the materials used to process functional semiconductor devices, and the performance of such devices. Therefore, it is important to any research and development endeavor involving semiconductor materials, and equally important to the commercial manufacturing of semiconductor devices, that the physical and chemical characteristics of the processed material are known and controlled at every stage of the fabrication procedures. For that reason, characterization of semiconductor materials and devices is an integral part of any semiconductor engineering effort (Schroeder, 2006; Runyan and Shattner, 1998; Seiler *et al.*, 1998). Equally important in this respect is monitoring of performance of the processes involved in the fabrication of semiconductor devices.

In this chapter, selected issues concerned with broadly understood semiconductor materials and process characterization/monitoring procedures are reviewed. The review focuses on the goals of materials characterization in semiconductor technology distinguishing between characterization of materials for the research and process development purposes, and the measurements aimed at the process monitoring and diagnostics.

Types of measurement techniques used in semiconductor characterization are identified, and the methods representing each class are briefly described, keeping in mind that to obtain complete qualitative and quantitative information regarding condition of the semiconductor material, methods from the different classes need to be used in the characterization procedure (Haight *et al.*, 2012).

The discussion in this chapter is concerned with materials and materials processing related effects and does not cover functional and parametric testing performed on the manufactured semiconductor devices. Those last topics are concerned with the aspects of semiconductor engineering not covered in this *Guide*.

The chapter is concluded with examples demonstrating how various characterization techniques are used to resolve selected challenges encountered in semiconductor device engineering.

6.2 Goals of Material and Process Characterization in Semiconductor Device Technology

The operation of semiconductor devices depends on the intricate physical interactions between outside stimuli (e.g. current, voltage, light, temperature), and semiconductor materials. Therefore, to assure predictable, controllable, and reproducible characteristics of the device, properties of the materials used to fabricate any given device must be known not only in terms of their inherent characteristics, but also in terms of the changes inflicted by the applied manufacturing processes used.

Science and engineering of materials characterization play an important role in semiconductor device technology in the areas identified below. As shown, there is a strong correlation between condition of the processed semiconductor material and measurable parameters defining its condition.

6.2.1 *Materials Research*

Before any material is included in semiconductor device fabrication, whether it is a semiconductor, conductor or insulator, it must be very well understood with respect to its effect on device performance as well as its changing basic characteristics under any given process conditions. An understanding of the material's usefulness in semiconductor device technology is based on the results of the fundamental studies of the properties of any given material which determines suitability of various materials for various device applications. For instance, Table 2.4 provides a summary of the device implications of selected physical characteristics of semiconductor materials. Similar considerations are applied to non-semiconductor materials, although different materials' characteristics are evaluated. Only in the case when properties of materials are well known and understood from the research perspective, any given material can be included in semiconductor device technology.

As an example of materials research effect in device technology, Fig. 6.1 shows decreasing minority carrier lifetime with increasing roughness of germanium wafer (Drummond *et al.*, 2011). What a curve plotted in this figure is telling us is that the parameters of processes performed on germanium wafer in the course of device fabrication must be researched and then selected keeping in mind possible roughening of its surface. Decrease of the minority carrier lifetime in the near surface region has an adverse effect on the performance of essentially any semiconductor device.

Materials research in semiconductor device engineering must go beyond semiconductor materials themselves. As the discussion in Chapter 3 has revealed, no functional semiconductor device can be constructed without insulators and conductors, mostly metals, incorporated into its structure. Therefore, materials research in semiconductor engineering can be concerned with essentially any type of solid regardless of its electrical conductivity.

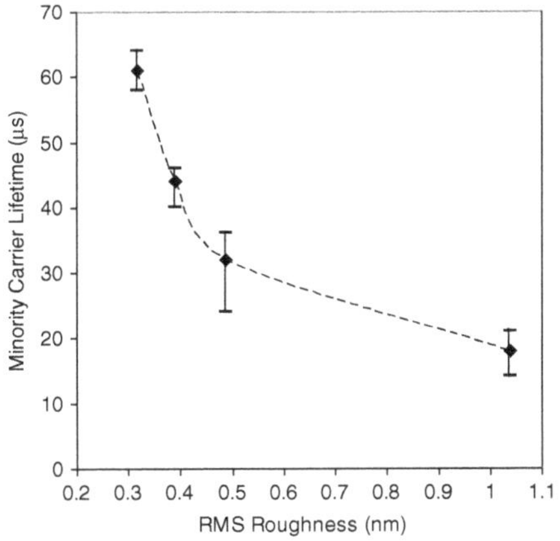

Fig. 6.1. Dependency of minority carrier lifetime in the near-surface of germanium wafer as a function of the roughness of its surface.

6.2.2 *Process Development*

Process development in semiconductor device materials characterization technology is an extension of the materials research as defined above into the device manufacturing environment. Development of the sequence of operations which convert raw semiconductor material into a functional device featuring desired specifications involves manipulation of the semiconductor properties using tools discussed in Chapter 4, and methods considered in Chapter 5. The effect of any operation on the properties of processed material needs to be known and understood. Thus, materials characterization in the context of a specific sequence of operations performed on the wafer is at the core of any process development effort.

Selection of material characteristics investigated at the process development stage depends on the device to be manufactured. For instance, in the case of the MOSFET discussed in Section 3.5.3, of importance are breakdown characteristics of the gate oxide (Wu *et al.*, 2001). For the sake of demonstration of the concept, Fig. 6.2 shows changing distribution of oxide breakdown events depending on silicon surface treatments applied before oxidation processes (*A* and *B* in the figure).

As oxide breakdown statistics shown in Fig. 6.2 demonstrate, superior oxide breakdown characteristics in the case of pre-oxidation process *A* are observed. Higher breakdown voltage and tighter distribution of breakdown events and in this case are an indication of less defective oxide.

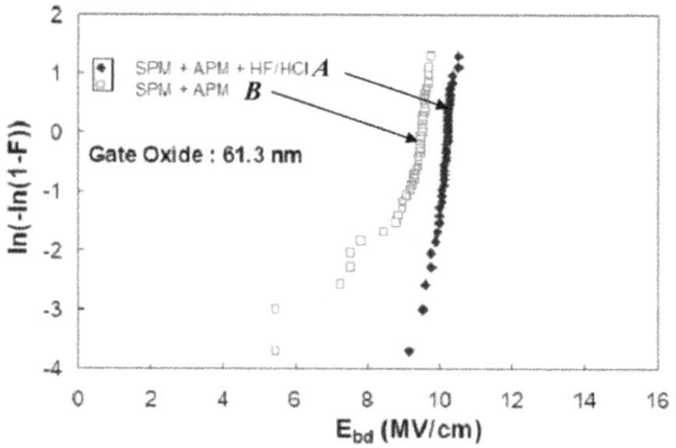

Fig. 6.2. Changes in the oxide breakdown statistics related to pre-oxidation surface processing recipes *A* and *B*.

Consequently, process *A* rather than process *B* is going to be used in commercial manufacturing of devices for which processes are developed.

As defined above both materials research and process development related semiconductor characterization are to be seen as the steps carried out in preparation of the commercial fabrication of semiconductor devices. Based on the results of process development, specific methods and tools used in the manufacture of a given device are determined. Choices in this regard have a direct impact on the way process monitoring operations are implemented.

6.2.3 *Process Monitoring*

It is essential to the outcome of the fabrication procedures to know what is the condition of the semiconductor wafer and structures formed on its surface at every stage of the device fabrication process. Thus, semiconductor manufacturing processes must be thoroughly controlled in terms of process parameters but also monitored in terms of the condition of the processed materials (Moyne *et al.*, 2001; May and Spanos, 2006). In contrast to materials research and process development considered earlier, process monitoring is implemented on the production line.

The goal of the process monitoring is to detect possible process malfunction as soon as it occurs so that line can be stopped and no additional wafers would be exposed to malfunctioning operations. Otherwise, with wafers getting larger and more expensive, cost of manufacturing processes increasing, and the number of operations performed on each wafer growing, losses resulting from any process malfunction could be very high.

Process monitoring is typically performed on the designated test wafers, strictly following established procedures. Test wafers are transferred for testing to the specialized laboratory and are not returned to the production lot. Such *off-line* process monitoring, schematically illustrated in Fig. 6.3(a), cannot detect process malfunctions in real-time, and thus, needs to be supplemented with *in-line* process monitoring involving wafers being processed, and the ambient in which wafers are processed (Fig. 6.3(b)).

Desired is an *in situ*, real-time process monitoring performed not on designated test wafers, but on each production wafer. This means that the method applied needs to be non-invasive in any way. In general, *in-line*, real-time process monitoring is better compatible with the single-wafer processes than with batch processes.

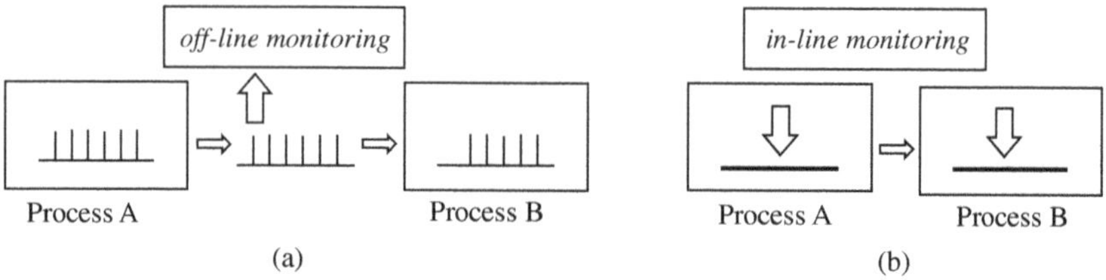

Fig. 6.3. (a) Off-line process monitoring on designated test wafers and (b) in-line process monitoring on the production wafers.

The use of the reflection high-energy electron diffraction (RHEED) instrumentation integrated with the MBE tool for the purposes of monitoring of the epitaxial growth (Fig. 5.14) is an example of this approach. On the other hand, the use of residual gas analyzers (RGA) to monitor composition of the gas ambient in the vacuum systems (Fig. 4.5) is an example of a process monitoring procedure not involving wafers being processed. Yet another class of process monitoring is exemplified by the *in-situ* film thickness measuring instruments installed in the PVD tools (Fig. 5.12). Typically, quartz crystal resonators are used for this purpose. Mounted inside the deposition chamber, crystal resonator reacts to the material deposition by changing its frequency of resonance. Upon calibration, information on both deposition rate and film thickness can be obtained this way.

Issues related to process monitoring in the semiconductor device fabrication are further considered later in this chapter.

6.2.4 *Process Diagnostics*

Any process malfunction unraveled by process monitoring needs to be diagnosed so that its cause is well understood, and hence, can be eliminated. The methods used in this case are also employed to optimize conditions of the manufacturing processes and are basically the same as those used in materials research and process development. Process diagnostics are typically carried out in the dedicated materials characterization laboratories, taking advantage of any available material characterization method that can shed light on the causes of process malfunction. For the sake of illustration of the role of process diagnostics, let's consider the MBE process monitored by RHEED apparatus as shown in Fig. 5.14. As soon as process malfunction is detected, manifesting itself, for instance, by structural faults in deposited epitaxial film detected by RHEED, the MBE tool would have to be stopped, and the cause of process malfunction determined during process diagnostics procedures. In the case considered here it could be a leak in the vacuum systems, malfunction of the vacuum pump or escaping control temperature of epitaxial film deposition. In this case diagnostics of the process malfunction is relatively straightforward. If the process would be allowed to continue and process malfunction detected based on the performance of the fabricated device, then the detection would involve time consuming multistep diagnostics procedure.

For this very reason it is essential that in the manufacture of semiconductor devices each process is monitored in real time. Due to the way they are applied, electrical measurements, if adequately implemented, can be very useful. See discussion later in this chapter for the additional considerations of this topic.

6.3 Methods of Material Characterization and Process Monitoring in Semiconductor Device Technology

The review of methods used in semiconductor materials and process characterization distinguishes spectroscopic, optical, and electrical techniques as well as often used methods which do not fall into any of these categories. There is a broad range of methods of semiconductor characterization in each category and the discussion of several of them in this section would not be compatible with the scope of this *Guide*. Therefore, only selected examples, seen as adequate representatives of the methods within each category, are considered below. It can be assumed that most of the methods reviewed are available in major industrial laboratories and research universities involved in advanced semiconductor research and development.

6.3.1 *Methods Based on X-ray, Electron and Ion Excitation*

Several techniques in this category operate based on the principles of spectroscopy. It relies on the determination of the spectra obtained as a result of solid material interactions with the excitation energy carried by short-wavelength electromagnetic radiation such as X-rays, or energy and electric charge carrying electrons, or ions.

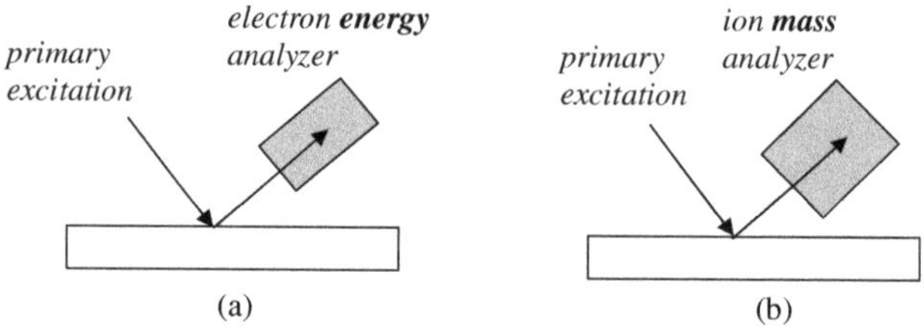

Fig. 6.4. (a) Origin of emitted electrons is based on the analysis of their energy and (b) in the case of sputtered ions their origin is determined based on their mass.

In the case primary excitation used is causing ejection of electrons from the irradiated solid, their origin, i.e. material they are emitted from, is based on the analysis of their energy using energy analyzer (Fig. 6.4(a)).

In the case of excitation energy causing sputtering of ions from the irradiated solid, their origin is determined based on their mass. Sputtered ions are captured by ion mass spectrometer which determines their mass and then, based on the established standards, their origin (Fig. 6.4(b)).

To ensure interactions of this nature are not disturbed, methods in this category need to be carried out in vacuum environment.

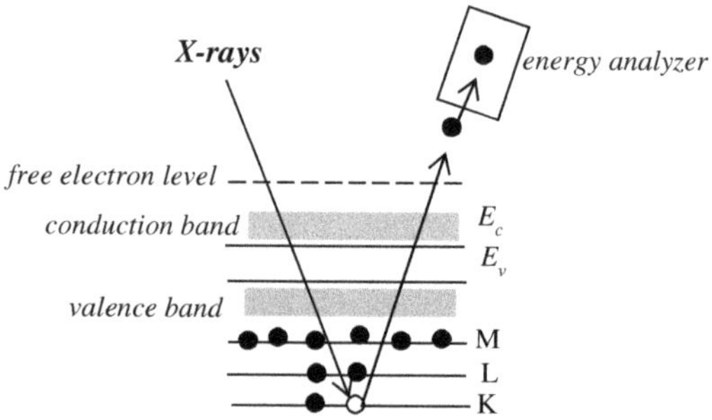

Fig. 6.5. Inter-level transitions upon which operation of X-ray photoelectron spectroscopy (XPS) is based.

X-ray Photoelectron Spectroscopy (XPS): As an example of method in the category considered, Fig. 6.5 illustrates working of the spectroscopic characterization technique using X-ray excitation which causes emission of photoelectrons energy of which is detected and analyzed by energy analyzer. The X-ray photoelectron spectroscopy (XPS), also known as electron spectroscopy for chemical analysis (ESCA), is commonly used to identify atomic and molecular species including oxygen, carbon, or fluorine (but not hydrogen) on the surface of the solids, including semiconductor materials.

Electron and ion spectroscopies: Other than X-ray-based class of spectroscopic methods applied in semiconductor characterization, involves techniques using beams of electrons or ions impinging on the surface and initiating desired interactions. In the latter case, the beam of ions, typically argon Ar^+, is bombarding the surface causing ejection of ions from the investigated material. Ejected ions are then captured by the mass spectrometer which determines mass, and then, based on the established standards, their origin (Fig. 6.4(b)).

Auger electron spectroscopy (AES): The AES is yet another technique often used in semiconductor characterization to determine chemical composition of the material, in its near-surface region in particular (Ruzyllo *et al.*, 1976). Most commonly, it uses electron beam excitation to initiate a two-stage process of ionization of an atom which is known as the Auger effect. The result is the emission of the secondary electrons referred to as Auger electrons of which energy spectra are used to determine the identity of the emitting atoms, as well as some information about their environment.

Secondary ion mass spectroscopy (SIMS): It is a method applied in semiconductor materials characterization based on ion mass spectroscopy. It allows determination of the composition of material leaving its surface severely damaged.

Depth profiling: A strength of methods such as listed above XPS, AES, and SIMS, is that in addition to the surface and near-surface analysis of material composition, they can also be used to determine in-depth distribution of selected atoms using procedure known as depth profiling (Fig. 6.6). It involves gradual ion sputtering of the material combined with surface analysis, for instance, AES (Fig. 6.6(a)), performed at predetermined time intervals while top layers are continuously sputter-etched. Depth profiling of the oxidized silicon (Fig. 6.6(a) and (b)), produces results as shown in Fig. 6.6(c).

Electron diffraction: A class of its own among material characterization methods using electron beams are based on electron diffraction. They are

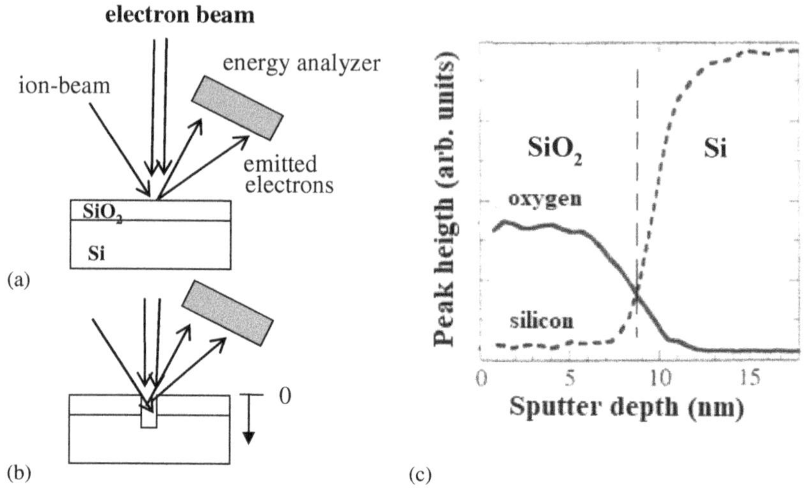

Fig. 6.6. Principles of depth profiling of oxidized silicon where (a) AES is combined with (b) gradual ion sputtering performing material analysis while oxide and then silicon are gradually removed, and (c) resulting in the profiles showing distribution of oxygen and silicon in the sample.

similar in nature to X-ray diffraction techniques mentioned earlier, except that electrons penetrate shorter distances in the solid than X-rays, and thus, can be effective in the determination of the lattice features of the geometrically confined 2D crystals. The monitoring of the growth of the ultra-thin epitaxial films as shown in Fig. 5.14, where reflection high-energy electron diffraction (RHEED) is an example of how electron diffraction techniques can be used in semiconductor technology. The choice between low-energy electron diffraction (LEED), high-energy electron diffraction (HEED), and RHEED depends on application.

Electron microscopy: A separate group of methods in semiconductor materials and devices characterization using electron excitation involves methods of electron microscopy. The most common in this group is scanning electron microscopy (SEM) which allows 3D visualization of the very fine features on the surface of the wafer including patterns created during device manufacturing.

Very useful variation of the electron microscopy is transmission electron microscopy (TEM) which is used to study multilayer material structures. Following tedious sample preparation procedures, TEM can distinguish between various materials in the multilayer structures with atomic-level resolution. For instance, using TEM, thickness of the few nm thick high-*k* dielectric film in the MOS gate stack can be determined (Fig. 2.7).

Infrared spectroscopy: Commonly used in materials research and engineering, a class of methods referred to as infrared (IR) spectroscopy also finds uses in semiconductor characterization. While methods listed above relied on the short wavelength UV and X-rays radiation, or ion and electron beams, infrared spectroscopy uses longer wavelengths infrared light to interact with a matter for the purposes of obtaining information regarding its properties. The types of interactions may involve absorption, emission, or reflection of infrared light, each delivering information regarding nature of chemical bonds regardless of whether matter investigated is a solid, liquid or gas. A variety of infrared spectroscopy delivering high-resolution data by identifying chemical bonds over the broad infrared absorption spectrum is known as Fourier Transform Infrared Spectroscopy (FTIR).

6.3.2 *Optical Methods*

A separate category of materials characterization methods used in semiconductor engineering is defined as *optical* because they use optical wavelengths from UV to visible to infrared, to interact with solids (Herman, 1996; Perkowitz, 2012). An important advantageous characteristic of the optical measurements is that they are non-destructive and non-invasive which means that they do affect measured materials in the negligible fashion.

Ellipsometry: Other than optical microscopy, which is the most obvious way to observe fine features of the sample, ellipsometry is a technique broadly used in semiconductor research and engineering to study thin-film materials which are transparent to light. In everyday semiconductor engineering applications ellipsometry is used to measure thickness of thin-film dielectrics. As Fig. 6.7 shows, ellipsometry detects a change in polarization of the short-wavelengths light upon its reflection from the

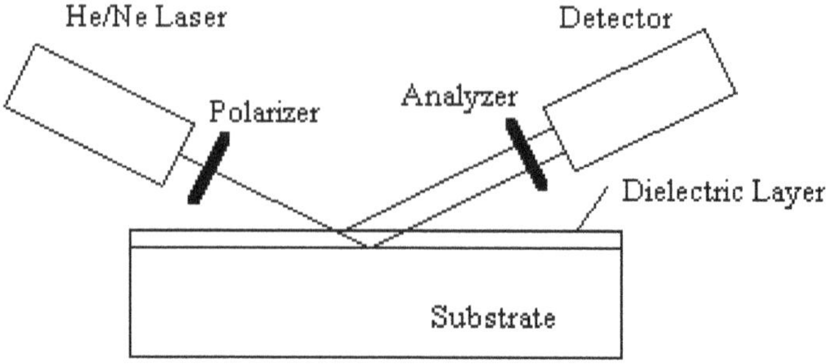

Fig. 6.7. Schematic illustration of the operation of a single wavelength ellipsometer.

air-measured material interface, and measured material-substrate interface. Characteristics of the film, including its thickness, are then determined by comparing them to the models established for any material system.

The advanced version of ellipsometry, known as spectroscopic ellipsometry, uses various wavelengths of the incident light and allows characterization of the complex, multilayer systems.

Infrared interferometry: Interferometry is a technique of the optical characterization of materials in which beams of the short-wavelength light originate from the same source are superimposed causing the effect of interference. It represents yet another commonly exploited material characterization method based on optical effects.

In semiconductor process monitoring applications, for instance, laser interferometry, capable of detecting changes of the refractive index of the irradiated material, is used in the end-point detection in dry etching operations. Laser interferometers are installed in the etching tools to signal completion of the etching of any given thin-film material, and thus, helping in assuring selectivity of etching processes.

Optical emission spectroscopy (OES): Alternative to laser interferometry method of end-point detection is OES. This technique detects the change in the wavelength of light emitted from plasma during etching. A shift in the emitted wavelengths follows the change in the composition of the gaseous ambient inside the etch tool at the point in time when the etched material is completely removed and etching of the underlying material is initiated.

Particle counting: Yet another example of the optical methods use in semiconductor process monitoring involves particle counting. As discussed in Section 4.6.1, particles adsorbed at the surface of the processed material are responsible for the major defects-causing effects in semiconductor technology, particularly in the case of devices featuring very tight geometries. In the particle counters the light shed on the surface of the wafer is scattered on the particles present there, and the resulting light points are detected and counted by the instrument. Advanced particle counters can detect and determine size of the particles in the nanometer range (Hattori, 1998).

Somewhat different methodologies are employed in the detection and measurements of airborne particles in the cleanroom air, as well as particles in deionized water and chemicals used in semiconductor manufacturing.

6.3.3 *Electrical Methods*

Unlike techniques of semiconductor materials characterization considered above, electrical methods reveal properties of the material which are the direct predictors of the performance of the device. For instance, while the causes of the reduced electron mobility in the processed semiconductor may not be immediately known, as it could be due to the structural defects, contamination, or other reasons, the adverse impact of the lower electron mobility on the device performance is certain. For that reason, electrical characterization of semiconductor materials providing direct quantitative information plays a special role in semiconductor device engineering. Parameters such as mobility of charge carriers, charge carrier lifetime, oxide breakdown voltage, as well as surface and interface charges between others, can provide information regarding performance and reproducibility of processes involved in semiconductor device fabrication.

To assure reliable determination of any given electrical parameter of semiconductor material, electrical communication between the material and the measuring circuit needs to be established. Such electrical communication can be realized in various ways schematically summarized in Fig. 6.8.

Selection of the method used to determine electrical characteristics of semiconductor sample depends on what part of the substrate, its bulk or its surface and near-surface region, needs to be electrically characterized.

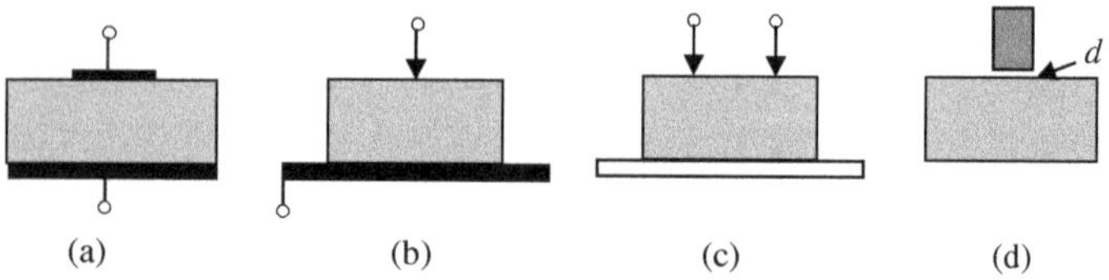

Fig. 6.8. Various ways of implementing electrical characterization of semiconductor materials including (a) permanent contact, (b) temporary contact, (c) lateral temporary contact, (d) non-contact electrical characterization.

In some cases, specifically in the devices which operation is based mostly on charge transport in the direction perpendicular to the surface, and thus bulk of the semiconductor sample plays a role in defining transport conditions, vertical test structures featuring permanent, deposited thin film metal contacts are used (Fig. 6.8(a)). The same measurements can be carried out using temporary contacts featuring sharp point contact or mercury probe to connect to the measured sample (Fig. 6.8(b)). In the case near-surface characteristics of semiconductor material need to be determined using electrical methods, current flow in a direction parallel to the surface can be implemented to effectively separate near-surface phenomena from those bulk related. Similarly to vertical structures, also in this case measurements can be performed using lateral temporary contacts (Fig. 6.8(c)) (Drummond, 2011). Alternatively, certain electrical characteristics of semiconductor materials can be derived using non-contact method under the conditions where surface illumination with light generating free charge carriers gives rise to the Surface Photovoltage (SPV) (Kamieniecki *et al.*, 1994). Detection of the characteristics of material in this case takes place *via* capacitive coupling over a very thin air gap d (Fig. 6.8(d)) in the range of 20 μm. The SPV signal reflecting condition of the illuminated surface and near-surface region is determined by measuring the displacement current. The SPV method described is non-invasive, which means it does not alter condition of the measured surface. The method can be made compatible with *in-line* process monitoring on product wafers in the case when RF-enhanced coupling, allowing separation d between the probe and monitored material in the range of millimeters (Fig. 6.8(d)), is applied instead of capacitive coupling (Kamieniecki, 2011).

It is quite common that the measurements of electrical characteristics of semiconductor test structures emphasize properties of the surface and near-surface region of the processed material (Roman *et al.*, 1998). The reason is the discussed earlier in this volume role of thin-films, surfaces, and interfaces in defining performance of the leading semiconductors devices including MOSFETs, and CMOS cells. Since the results of these measurements are also a good litmus test of the material characteristics defining performance of the majority of other than the MOSFET semiconductor devices, the MOS capacitors are common test structure in semiconductor characterization. The electrical parameters obtained this way can provide information of the broad range of semiconductor material characteristics.

Various electronic properties of semiconductors materials which affect performance of the MOSFET can be measured using MOS capacitors. This way, the need to process complete transistors for the material characterization, or process diagnostics purposes, can be avoided. Figure 6.9 identifies types of measurements that can be performed as a function of applied voltage using MOS capacitors and lists physical characteristics of the material system which can be obtained from such measurements. In addition to capacitance (*C*) and current (*I*) measurements shown in Fig. 6.9, measurements of conductance as a function of signal frequency can be performed to obtain more detailed information regarding surface and interface characteristics.

Overall, electrical characterization of semiconductors is very well established in semiconductor device engineering and plays pivotal role in the broad range of characterization applications. Over the years, many methods and methodologies of electrical characterization were subjected to changes in response to the needs of nanometer scale device geometries,

C, capacitance (C-V)

- surface barrier height
- effective charge density
- surface state density
- dopant concentration at the surface

I, current (I-V)

- gate oxide breakdown voltage
- charge to breakdown
- oxide leakage current

Fig. 6.9. Examples of parameters that can be determined from the capacitance–voltage and current–voltage measurements of the MOS capacitor.

ultra-thin films, and 3D device features. However, principles underlying electrical characterization of semiconductor materials remain unchanged.

6.3.4 *Other Methods*

In addition to the methods discussed above, there is a group of semiconductor materials characterization methods which are based on physical and chemical phenomena, but do fall into the categories considered. Two examples of such methods are introduced in the following.

Atomic force microscopy (AFM): Atomic Force Microscopy is a method of choice when there is a need for the visualization and measurement of atomic-scale surface roughness. Roughness is a physical property of single-crystal semiconductor surfaces with an effect on the manufacturing yield and performance of devices formed on such surfaces. It is therefore essential that surface roughness is controlled throughout the device manufacturing process. The method of AFM is a method of choice when there is a need for investigation of atomic-scale surface roughness. The AFM characterization can be carried out in air or in vacuum with no special preparation of the sample needed. Commonly used to define surface roughness is calculated root mean square (RMS) of the measured microscopic peaks and valleys appearing on the surface.

To exemplify the results of AFM measurements, Fig. 6.10 shows 2D and 3D view of silicon surfaces showing AFM visualization of the surface condition during silicon device fabrication (Drummond *et al.*, 2011). As seen in this figure, AFM allows monitoring of the surface condition in terms of roughness during device fabrication.

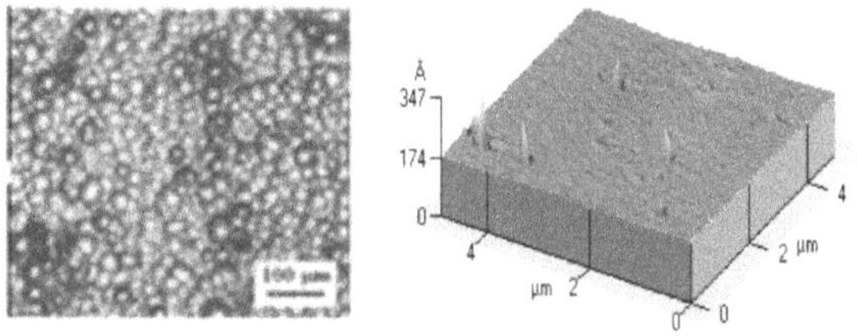

(a) (b)

Fig. 6.10. (a) 2D AFM image of the surface of multicrystalline Si damaged by excessive etching and (b) 3D AFM image of epitaxial Si in which surface defects of the substrate are reproduced.

The AFM is the most used method to precisely measure and quantify surface roughness in semiconductor technology and as such is commonly available in semiconductor research laboratories and manufacturing facilities.

Wetting angle measurements: Measurements of the wetting (contact) angle is a material characterization technique which is straightforward, yet able to instantaneously provide relevant information regarding coverage (termination) of the solid surfaces and as such is useful in process control applications.

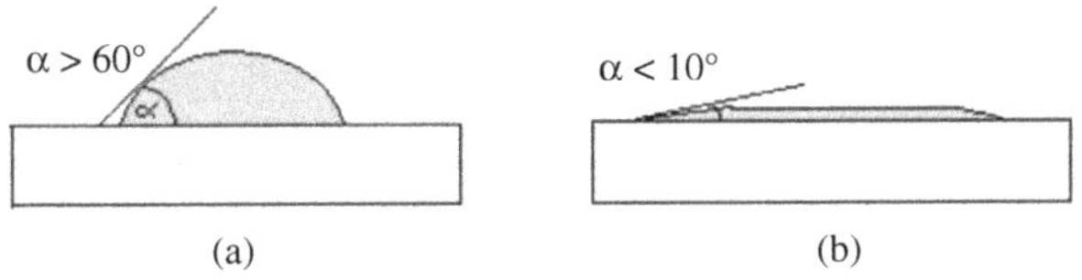

Fig. 6.11. Wetting (contact) angle α and water on (a) hydrophobic surface and (b) hydrophilic surface.

Using simple instrumentation in the laboratory ambient air, measurements of the wetting angle distinguish between surfaces featuring different surface energies, including the difference between hydrophobic (wetting angle approaching 90°) surfaces and hydrophilic (wetting angle α close 0°) surfaces. In the former case water is bidding on the surface (Fig. 6.11(a)) while in the latter water is wetting the surface (Fig. 6.11(b)).

An advantageous feature of the contact angle measurements is that they can be implemented on the as-processed surfaces without any prior preparatory treatments. This makes contact angle measurements potentially useful in process monitoring applications.

6.4 Examples of Material and Process Characterization in Semiconductor Device Technology

In this section, examples of material and process characterization methods used in semiconductor device technology are considered following identified in Section 6.2 areas in which such methods may be applied.

Regarding development of the process concerned with fin structures (see Fig. 3.17) it is good to know correlation between width of the fin and

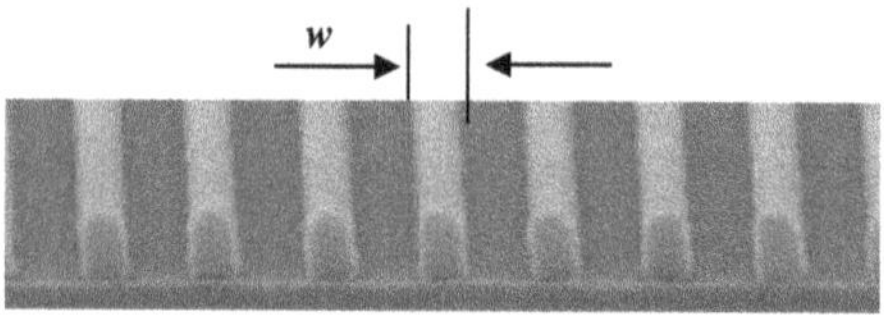

Fig. 6.12. Electron microscopy image of fins featuring width w etched in silicon.

electronic properties of silicon in which fin is formed. This is because material properties of interest are likely to be different in planar and 3D configurations, for instance. In addition, when an aggressive nanometer confinement is involved, materials to be used may change their properties in meaningful fashion as compared to the same material in geometrically relaxed configuration.

As far as the impact of device architecture on materials properties, knowledge of the properties of materials affected needs to be complete and well documented. As an example, Fig. 6.12 shows device architecture based on fins formed on the Si surface. As the measurements of minority carrier lifetime performed on the samples featuring different fin's width w have revealed (Wakil *et al.*, 2012), structures featuring wider fins featured higher minority carrier lifetime.

The other example of the use of combination of methods listed above in process development is shown in Fig. 6.13. It concerns often encountered in device processing challenges related to the properties of the ohmic contacts. The key issue here is control of the residual interfacial oxide between metal and semiconductor (see Fig. 5.31(b)) which has an adverse effect on the contact properties. Using metal contact to silicon as an example, Fig. 6.13(a) shows XPS results indicating presence of the oxide phase on the surface (see high oxygen, O1s peak). When prior to metal deposition residual oxide is etched off using HF (1):H_2O (100) solution then O1s peak is reduced to the value indicating no residual oxide on the surface (Fig. 6.13(b)).

To confirm validity of the effect of HF treatments indicated by XPS results on actual ohmic contact characteristics, actual devices need to be processed as a part of process development effort. Figure 6.13(c) shows in a qualitative fashion how current–voltage (*I*–*V*) characteristics of the contact change toward lower series resistance when silicon surface prior to metal deposition is subjected to HF-treatment. The fact that the removal of the residual oxide was responsible for the improved *I*–*V* characteristics of the contact supports results of XPS analysis showing reduced oxygen

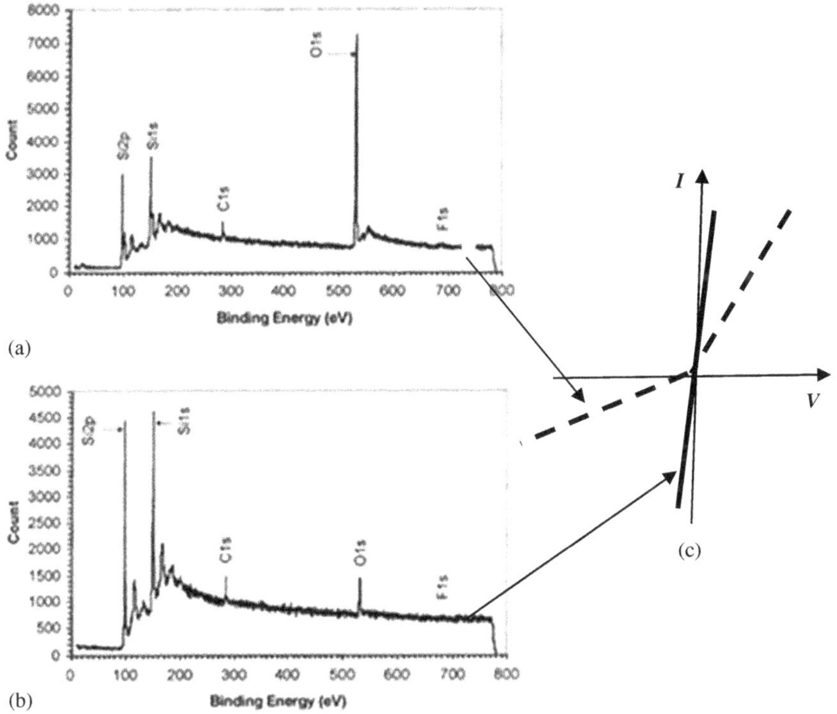

Fig. 6.13. (a) XPS spectrum of Si surface before HF treatment, (b) XPS spectrum of the same surface after HF treatment, and (c) *I–V* characteristics of the metal-Si contact before and after HF treatment.

peak and enhanced Si signal (Fig. 6.13(b)). Therefore, it can be assumed that HF etching performs according to expectation and pre-metal deposition surface treatment can be included in the device fabrication process.

The point to be made here is that material characterization, best if supported by electrical measurements, is an integral part of any semiconductor device manufacturing process development. The methods of semiconductor material and process characterization at our disposal, when used in combination, can be of assistance in the resolution of essentially any materials related challenge encountered in semiconductor device engineering.

Another example of material characterization in the process development is concerned with sapphire substrates which are used in silicon-on-sapphire (SOS) technology discussed in Section 2.8.2. Important part of

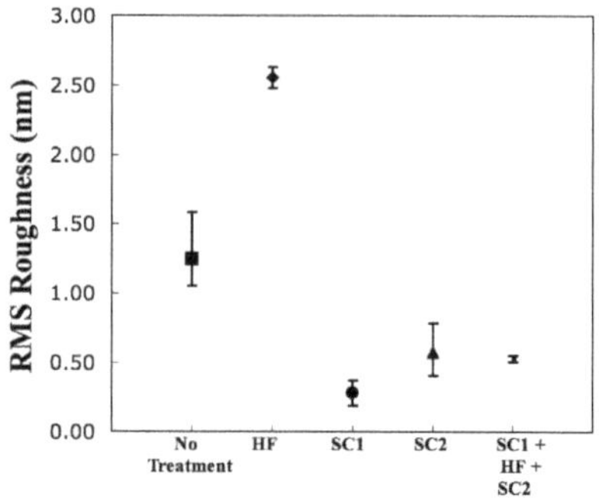

Fig. 6.14. The effect of chemistry used in preparation of sapphire surfaces prior to epitaxial layer silicon on the roughness of sapphire surface.

the process in this case is the appropriate preparation of sapphire surface prior to deposition of thin-film silicon with surface roughness of sapphire surface having effect on the process (Kirby *et al.*, 2007). Testing in this application wet chemistries used in silicon surface cleaning have an effect on the roughness of sapphire surface as seen in Fig. 6.14.

The results shown in Fig. 6.14 indicate adverse impact of HF-based solution on sapphire surface. However, when used in combination with SC1 and SC2 treatments, (SC1+HF+SC2) HF etches off oxide residues from sapphire surface without roughing it. Based on the results obtained, the process of sapphire surface treatments prior silicon deposit can be developed for the fabrication of SOS substrates.

Next in this discussion is an overview of the ways material characterization techniques can be used in monitoring processes, or in process control, involved in fabrication of semiconductor devices. The goal is detection of process malfunctions in real time. As Fig. 6.3 indicated, process monitoring procedures can be implemented *off-line* or as *in-line* characterization methods performed in real time on the production wafers. To be compatible with production wafers the methods used in *in-line* process monitoring cannot alter in any way characteristics of the measured material. In contrast, *off-line* characterization is performed on the test wafers which once removed for testing, cannot be returned to the production lot.

To meet this requirement, the method used in in-line process monitoring needs to be non-invasive to the processed wafer while providing relevant information regarding performance of the process. Compatible with

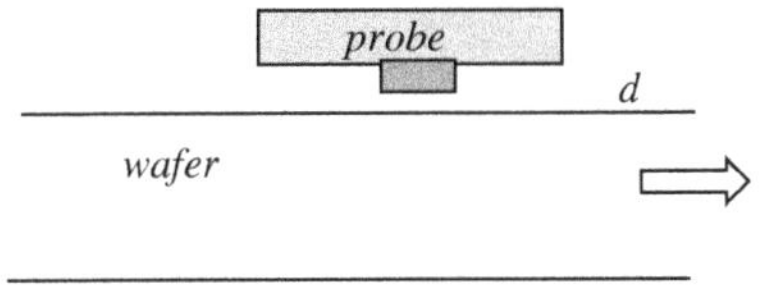

Fig. 6.15. Process monitoring using non-contact probe allowing measurement of surface charge on Si surface which changes depending on the outcome of any given process.

this application are methods of non-contact electrical measurements (Fig. 6.8(d)), for instance, measuring surface charge on the wafer moving from one tool to another in the cluster (Fig. 4.16). Figure 6.15 demonstrates the use of Surface Charge Profiling (SCP) method to identify various features created on Si surface, for instance, as a result of process malfunction.

Advantageous characteristics of measurements based on the determination of surface charge is sensitivity of this parameter to essentially any change is surface characteristics. As discussed earlier, the method is best compatible with in-line process monitoring on product wafers in the case when RF-enhanced coupling allowing separation between the probe and monitored material in the range of mm (*d* in Fig. 6.15) is implemented (Kamieniecki, 2011).

While detecting malfunction of the monitored process, *in-line* methodology considered above may not provide information based on which conclusion regarding nature of the detected process malfunction can be drawn. Information in this regard is needed to prevent the same interruption of the device fabrication procedure in the future by modifying tools and method upon which process is based. For those reasons, detection of the process malfunction needs to be followed *off-line* by the process diagnostic procedures. In *off-line* laboratory facilities, which are commonly available in semiconductor devices manufacturing sites, tools implementing at least some of the methods discussed in this *Guide* are typically accessible, and hence, can be used in process diagnostics. The goal of these measurements is to determine exact nature of the process failure originally detected during *in-line* process monitoring performed on the product wafers. The results of this analysis will allow process optimization to prevent similar process failures in the future.

As the discussion in this chapter demonstrated, semiconductor material and process characterization is an integral component of semiconductor device engineering. While hands-on experience with methods discussed require specialized training, general understanding of the nature of the procedures involved and the ways of their implementation is a measure of readiness to be involved in activities related to semiconductor device technology.

Review Questions

6.1. Why are semiconductor materials and process characterization methods needed in semiconductor device technology?
6.2. Consider implementation of semiconductor materials characterization methods in process development.
6.3. Explain the difference between *off-line* and *in-line* process monitoring.
6.4. Give an example of the X-ray based methods in material characterization in semiconductor device technology.
6.5. Explain the process of depth profiling using appropriate illustrations.
6.6. Explain principle of operation of a single wavelength ellipsometer.
6.7. What material characteristics can be determined based on the electrical measurements of MOS capacitors?
6.8. What information regarding condition of semiconductor surface can be obtained from wetting angle measurements?
6.9. How XPS characterization is used in the development of ohmic contact processing in silicon device technology?
6.10. How noninvasive electrical characterization of semiconductor surface can be implemented?

Bibliography

Alcacer, L. (2018). *Electronic Structure of Organic Semiconductors: Polymers and Small Molecules*. Morgan and Claypool.

Aoki, H. and Dresselhaus, M. S. (Eds.). (2014). *Physics of Graphene*. Springer Verlag.

Arora, A., Drummond, P. J., and Ruzyllo, J. (2016). Electrical characterization of silicon on insulator wafers by photoconductance decay (PCD) method. *ECS Journal of Solid-State Science and Technology*, **5**(4), 3069.

Asahi, H. and Horikoshi, Y. (2019). *Molecular Beam Epitaxy: Materials and Applications for Electronics and Optoelectronics*. Wiley.

Baca, A. C. and Ashby, C. I. H. (2005). *Fabrication of GaAs Devices*. Institution of Electrical Engineers.

Baker, R. J. (2019). *CMOS Circuit Design, Layout and Simulation* (4th ed.). J. Wiley and Sons.

Bakhtari, K., Guldiken, A. O., Busnaina, A. A., and Park, J. (2006). Experimental and analytical study of submicrometer particle removal from deep trenches. *Journal of the Electrochemical Society*, **153**(9).

Baklanov, M., Green, M., and Maex, K. (Eds.). (2007). *Dielectric Films for Advanced Microelectronics*. J. Wiley and Sons.

Bandyopadhyay, S. and Cahay, M. (2016). *Introduction to Spintronics* (2nd ed.). CRC Press.

Bardeen, J. and Brattain, W. H. (1949). Physical principles involved in transistor action. *Physical Review*, **75**(8), 1208.

Beck, R., Patyra, M., Ruzyllo, J., and Jakubowski, A. (1980). Oxygen pressure as a parameter in the DC plasma anodization of silicon. *Thin Solid Films*, **67**(2), 261–264.

Berger, L. I. (1997). *Semiconductor Materials*. CRC Press.

Bisquert, J. (2017). *The Physics of Solar Cells*. CRC Press.

Bonca, J. and Kruchinin, S. (Eds.). (2010). *Physical Properties of Nanosystems*. Springer-Verlag.
Brophy, J. J. (1964). *Semiconductor Devices*. McGraw-Hill.
Burghartz, J. N. (Ed.). (2013). *State-of-the-Art Electron Devices*. J. Wiley and Sons.
Campbell, S. A. (2020). *Fabrication Engineering at the Micro- and Nanoscale*. Oxford University Press.
Chambers, S. A. (Ed.). (2025). *Handbook of Molecular Beam Epitaxy of Oxide Materials*. World Scientific Publishing Co.
Chang, K., Shanmugasundaram, K., Shallenberger, J., and Ruzyllo, J. (2007). Studies of Hf(Si,O) dielectrics for metal–oxide–semiconductor applications. *Thin Solid Films*, **515**(7–8), 3802–3805.
Choy, K. L. (2019). *Chemical Vapour Deposition (CVD): Advances, Technology and Applications*. CRC Press.
Clifford, T. (1998). *Fundamentals of Supercritical Fluids*. Oxford Academic.
Cooke, M. J. (1990). *Semiconductor Devices*. Prentice Hall.
Drummond, P. J., Bhatia, D., Kshirsagar, A., Ramani, S., and Ruzyllo, J. (2011). Studies of photoconductance decay method for characterization of near-surface electrical properties of semiconductors. *Thin Solid Films*, **519**(22), 7621.
Enderlein, R. and Horing, N. J. (1997). *Fundamentals of Semiconductor Physics and Devices*. World Scientific.
Erdamar, M., Shanmugasundaram, K., Roman, P., Chang, K., Klimkiewicz, M., and Ruzyllo, J. (2008). Deep lateral anhydrous HF/methanol etching for MEMS release processing. *Journal of Micro- and Nanolithography, MEMS and MOEMS*, **7**(3), 1.
Ferry, D. K. (2013). *Semiconductors: Bonds and Bands*. Institute of Physics.
Fischetti, M. V. and Vandenberghe, W. G. (2016). *Advanced Physics of Electron Transport in Semiconductors and Nanostructures*. Springer-Verlag.
Fornari, R. (2018). *Single Crystals of Electronic Materials: Growth and Properties*. Woodhead Publishing.
Geng, H. (2018). *Semiconductor Manufacturing Handbook* (2nd ed.). McGraw Hill.
Ghandi, S. K. (1983). *VLSI Fabrication Principles, Silicon and Gallium Arsenide*. J. Wiley and Sons.
Grahn, H. T. (1991). *Introduction to Semiconductor Physics*. World Scientific.
Grot, S. A., Hatfield, C. W., Gildenblat, G. Sh., Badzian, A. R., and Badzian, T. (1991). Electrical properties of homoepitaxial diamond films. *Applied Physics Letters*, **58**(14), 1542.
Grove, A. S. (1967). *Physics and Technology of Semiconductor Devices*. J. Wiley and Sons.
Grundmann, M. (2006). *Physics of Semiconductors*. Springer-Verlag.
Haight, R., Ross, F. M., and Hannon, J. B. (Eds.). (2012). *Handbook of Instrumentation and Techniques for Semiconductor Nanostructure Characterization*. World Scientific.

Hattori, T. (Ed.). (1998). *Ultraclean Surface Processing of Silicon Wafers*. Springer-Verlag.

Hattori, T. (1998). Detection and analysis of particles in production lines. In T. Hattori (Ed.), *Ultra Clean Surface Processing of Silicon Wafers* (pp. 243–258). Springer-Verlag.

Herman, I. P. (1996). *Optical Diagnostics for Thin Film Processing*. Academic Press.

Heyns, M., Mertens, P., and Ruzyllo, J. (1999). Advanced wet and dry cleaning coming together for next generation. *Solid State Technology*, **42**(3), 37.

Hirano, H., Sato, K., Osaka, T., Kuniyasu, H., and Hattori, T. (2006). Damage-free ultradiluted HF/nitrogen jet spray cleaning for particle removal with minimal silicon and oxide loss. *Electrochemical and Solid-State Letters*, **9**(2), G62–G65.

Hoff, A. M. and Ruzyllo, J. (1988). Atomic oxygen and thermal oxidation of silicon. *Applied Physics Letters*, **52**(15), 1264–1266.

Hossain, S., Pantano, C., and Ruzyllo, J. (1990). Removal of surface organic contaminants during thermal oxidation of silicon. *Journal of the Electrochemical Society*, **137**(10), 3287.

Hu, C. C. (2009). *Modern Semiconductor Devices for Integrated Circuits*. Pearson.

Hull, R. (Ed.). (1999). *Properties of Crystalline Silicon*. INSPEC Publications.

Hwang, D. K., Torek, K., and Ruzyllo, J. (1996). Slight etching of silicon to control post-RIE damage. In K. P. Cheung, M. Nakamura, and C. T. Gabriel (Eds.), *Proceedings of the First International Symposium on Plasma Process-Induced Damage*, May 13–14, Santa Clara, CA (pp. 137–140).

Irene, E. A. (2008). *Surfaces, Interfaces and Thin Films for Microelectronics*. J. Wiley and Sons.

Jackson, T., Mourey, D. A., and Zhao, D. (2011). *Plasma enhanced atomic layer deposition process* (U.S. Patent Application No. 12/872,040).

Jaeger, R. C. (2001). *Introduction to Microelectronic Fabrication* (2nd ed.). Pearson.

Jousten, K. (2016). *Handbook of Vacuum Technology*. J. Wiley and Sons.

Kaariainen, T., Cameron, D., Kaariainen, M.-L., and Sherman, A. (2013). *Atomic Layer Deposition* (2nd ed.). Wiley-Scrivener.

Kamieniecki, E., Roman, P., Hwang, D., and Ruzyllo, J. (1994). A new method for in-line, real-time monitoring of wafer cleaning operations. *Proceedings of the Electrochemical Society*, 189–195.

Kamieniecki, E. (2011). *Electrical characterization of semiconductor material* (U.S. Patent No. 7,898,280 B2). U.S. Patent and Trademark Office.

Kasap, S. O. (2006). *Principles of Electronic Materials and Devices* (3rd ed.). McGraw Hill.

Kern, W. and Puotinen, D. A. (1970). Cleaning solutions based on hydrogen peroxide for use in silicon semiconductor technology. *RCA Review*, **31**, 187.

Kirby, K., Shanmugasundaram, K., and Ruzyllo, J. (2007). Interactions of sapphire surfaces with standard cleaning solutions. *ECS Transactions*, **11**(2), 343.

Knotter, D. M., de Gendt, S., Mertens, P. W., and Heyns, M. M. (2000). Silicon surface roughening mechanisms in ammonia hydrogen peroxide mixtures. *Journal of the Electrochemical Society*, **147**, 736.

Kozicki, M., Hoenig, S. A., and Robinson, P. J. (1991). *Cleanrooms: Facilities and Practices*. Van Nostrand Reinhold.

Kshirsagar, A., Jiang, Z., Pickering, S., Xu, J., and Ruzyllo, J. (2013). Formation of photo-luminescent patterns on paper using nanocrystalline quantum dots ink and mist deposition. *ECS Journal of Solid State Science and Technology*, **2**(5), R87.

Kuo, W. and Kim, T. (1999). An overview of manufacturing yield and reliability modeling for semiconductor products. *Proceedings of the IEEE*, **87**, 1329.

Lau, J. H. (2021). *Semiconductor Advanced Packaging*. Springer.

Le Roux, V., Machicoane, G., Kerdiles, S., Laffitte, R., Bechu, N., Vallier, L., Borsoni, G., Korwin-Pawlowski, M., Roman, P., Wu, C.-T., and Ruzyllo, J. (2004). Etching of silicon native oxide using ultra-slow multicharged Ar^{q+} ions. *Journal of the Electrochemical Society*, **151**(1), G76.

Levinshtein, S., Rumyantsev, M., and Schur, M. (Eds.). (1999). *Handbook Series on Semiconductor Parameters*. World Scientific.

Lian, Y. (2022). *Semiconductor Microchips and Fabrication*. Wiley.

Liang, Y. C., Samudra, G. S., and Huang, C. F. (2017). *Power Microelectronics: Device and Process Technologies* (2nd ed.). World Scientific.

Lilienfeld, J. E. (1930). *Method and apparatus for controlling electric currents* (U.S. Patent No. 1,745,175).

Liu, C. (2011). *Foundations of MEMS* (2nd ed.). Pearson.

Lukasiak, L., Roman, P., Jakubowski, A., and Ruzyllo, J. (2001). Analysis of surface and interface charge interactions in SOI substrates. *Solid-State Electronics*, **45**, 95.

Lundstrom, M. (2000). *Fundamentals of Carrier Transport* (2nd ed.). Cambridge University Press.

Luther, B. P., Ruzyllo, J., and Miller, D. I. (1993). Nearly isotropic etching of 6H SiC in NF_3 and O_2 using remote plasma. *Applied Physics Letters*, **63**(2), 171.

Mack, C. (2007). *Fundamental Principles of Optical Lithography: The Science of Microfabrication*. J. Wiley and Sons.

Machlin, E. S. (2010). *Materials Science in Microelectronics I: The Relationships Between Thin Film Processing and Structure*. Elsevier Science.

Madelung, O. (2004). *Semiconductors: Data Handbook* (3rd ed.). Springer Verlag.

Majkusiak, B. (1998). Experimental and theoretical study of the current-voltage characteristics of the MISIM tunnel transistor. *IEEE Transactions on Electron Devices*, **45**(9), 1903.

Maly, W. (1987). *Atlas of IC Technologies*. Benjamin/Cummings Publishing Co.

May, G. S. and Spanos, C. J. (2006). *Semiconductor Manufacturing and Process Control*. J. Wiley and Sons.

Mayer, A. and Schwartzman, S. (1979). Megasonic cleaning: A new cleaning and drying system for use in semiconductor processing. *Journal of Electronic Materials*, **8**, 855.

McGuire, G. E. (1989). *Characterization of Semiconductor Materials, Principles and Methods* (Vol. 1). Noyes Publications.

McGuire, G. E. and Strausser, Y. S. (2010). *Characterization in Compound Semiconductor Processing*. Momentum Press.

Meuris, M., Verhaverbeke, S., Mertens, P. W., Schmidt, H. F., Rotondaro, A. L. P., Heyns, M. M., and Philipossian, A. (1994). A new cleaning concept for particle and metal removal on Si surfaces. *Proceedings of the Electrochemical Society*, PV 94–07, 15–25.

Mouche, L. and Tardif, F. (1995). Mechanisms of particle removal from silicon wafer surface in wet chemical cleaning process. *Journal of the Electrochemical Society*, **141**, 1684.

Moyne, J., del Castillo, E., and Hurwitz, A. M. (Eds.). (2001). *Run-to-Run Control in Semiconductor Manufacturing*. CRC Press.

Mumbauer, P., Chang, K., Mahoney, W., Lee, D.-O., Shanmugasundaram, K., Roman, P., Brubaker, M., Grant, R., and Ruzyllo, J. (2004). Mist deposition in semiconductor device manufacturing. *Semiconductor International*, **27**(12), 75–80.

Neamen, D. A. (2011). *Semiconductor Physics and Devices* (4th ed.). McGraw Hill.

Nishi, Y. and Doering, R. (Eds.). (2008). *Handbook on Semiconductor Manufacturing Technology* (2nd ed.). CRC Press.

Oliver, M. R. (Ed.). (2004). *Chemical-Mechanical Planarization of Semiconductor Materials*. Springer-Verlag.

Perkowitz, S. (2012). *Optical Characterization of Semiconductors*. Academic Press.

Pfitzner, A. (2011). Vertical-slit field-effect transistor (VeSFET) – design, space exploration and DC model. *Proceedings of MIXDES 2011*, 151.

Pierret, R. F. (1996). *Semiconductor Device Fundamentals*. Addison-Wesley.

Plummer, J. D., Deal, M., and Griffin, P. D. (2008). *Silicon VLSI Technology: Fundamentals, Practice, and Modeling*. Prentice Hall.

Pohl, U. W. (2020). *Epitaxy of Semiconductors*. Springer.

Price, S. C., Shanmugasundaram, K., Ramani, S., Zhu, T., Zhang, F., Xu, J., Mohney, S. E., Zhang, Q., Kshirsagar, A., and Ruzyllo, J. (2009). Studies of mist deposition for the formation of quantum dots CdSe films. *Semiconductor Science and Technology*, **24**(10), 105024.

Ravi, M., Chen, W., Thompson, P., Bottoms, W., Helmy, A. S., Chan, B., Swaminathan, M., Shalf, J., Bailey, C., Schutt-Aine, J., Subbarayan, S., Iyer, S. S., Narumanchi, S., and Mongia, R. (2026). The heterogeneous integration of electronic components. *Nature Reviews Electrical Engineering*, **3**, 254–263.

Reinhardt, K. A. and Kern, W. (Eds.). (2018). *Handbook of Silicon Wafer Cleaning Technology* (3rd ed.). William Andrew.

Ren, F. and Pearton, S. J. (Eds.). (2017). *Semiconductor Based Sensors*. World Scientific.

Ridley, R., Brozek, T., and Ruzyllo, J. (1995). The effect of carbon in silicon on oxidation kinetics and dioxide integrity. *ULSI Science and Technology, The Electrochemical Society*, 95-5, 443–463.

Rockett, A. (2008). *The Materials Science of Semiconductors*. Springer-Verlag.

Roman, P., Staffa, J., Fakhouri, S., Brubaker, M., Torek, K., Kamieniecki, E., and Ruzyllo, J. (1998). Surface dopant concentration monitoring using noncontact surface charge profiling. *Journal of Applied Physics*, **83**(4), 2297–2300.

Runyan, W. R. and Bean, K. E. (1990). *Semiconductor Integrated Circuit Processing Technology*. Addison-Wesley Publishing Co.

Runyan, W. R. and Shattner, T. J. (1998). *Semiconductor Measurements and Instrumentation* (2nd ed.). McGraw-Hill.

Ruzyllo, J., Shiota, I., Miyamoto, N., and Nishizawa, J. (1976). Low temperature oxidation of silicon studied by photosensitive ESR and Auger Electron Spectroscopy. *Journal of the Electrochemical Society*, **123**, 26.

Ruzyllo, J. (1980). Lateral MIS tunnel transistor. *IEEE Electron Device Letters*, **EDL-1**, 197.

Ruzyllo, J. (1998). New semiconductor materials. In S. K. Majumdar, R. E. Tressler, and E. W. Miller (Eds.), *The Era of Materials*. The Pennsylvania Academy of Science.

Ruzyllo, J. (2014). Assessment of the progress in gas-phase cleaning of silicon surfaces. *ECS Journal of Solid State Science and Technology*, **3**(1), N3060.

Ruzyllo, J. (2016). *Semiconductor Glossary*. World Scientific Publishing Co.

Ruzyllo, J. (2025). *Guide to Characteristics and Characterization of Semiconductor Surfaces*. World Scientific Publishing Co.

Sabba, D. (2017). *Organic Semiconductors*. Alcer Press.

Sabeeh, A. H., Brigeman, A. N., and Ruzyllo, J. (2019). Performance of single-crystal silicon solar cells with mist-deposited nanocrystalline quantum dot downshifting films. *IEEE Journal of Photovoltaics*, **9**(4), 1006.

Schroeder, D. K. (2006). *Semiconductor Material and Device Characterization* (3rd ed.). John Wiley and Sons.

Seiler, D. G., Diebold, A. C., Bullis, W. M., Shaffner, T. J., McDonald, R., and Walters, E. J. (Eds.). (1998). *Characterization and Metrology for ULSI Technology*. American Institute of Physics.

Sesham, K. (Ed.). (2012). *Handbook of Thin Film Deposition* (2nd ed.). William Andrew.

Shön, H. (2015). *Handbook of Purified Gases*. Springer-Verlag.

Shul, R. J. and Pearton, S. J. (Eds.). (2000). *Handbook of Advanced Plasma Processing Techniques*. Springer-Verlag.

Singh, J. (2007). *Electronic and Optoelectronic Properties of Semiconductor Structures*. Cambridge University Press.

Smith, D. L. (1995). *Thin-Film Deposition*. McGraw Hill.

Staffa, J., Hwang, D., Luther, B., Grant, R., and Ruzyllo, J. (1995). Temperature dependence of the etch rate and selectivity of silicon nitride over silicon dioxide in remote plasma NF_3/Cl_2. *Applied Physics Letters*, **67**(13), 28–32.

Stalliga, P. (2009). *Electrical Characterization of Organic Electronic Materials and Devices*. J. Wiley and Sons.

Streetman, B. G. and Banerjee, S. K. (1998). *Solid State Electronic Devices* (6th ed.). Prentice Hall.

Swaminathan, M. (2025). *Center for Heterogeneous Integration of Micro Electronic Systems (CHIMES)*.

Sze, S. M. (1985). *Semiconductor Devices Physics and Technology*. J. Wiley and Sons.

Sze, S. M. (Ed.). (1990). *High-Speed Semiconductor Devices*. J. Wiley and Sons.

Sze, S. M. (Ed.). (1998). *Modern Semiconductor Device Physics*. J. Wiley and Sons.

Sze, S. M., Li, Y., and Ng, K. K. (2021). *Physics of Semiconductor Devices* (4th ed.). J. Wiley and Sons.

Tan, Z., Zhang, F., Zhu, T., Xu, J., Wang, A. Y., Dixon, J. D., Li, L., Zhang, Q., Mohney, S. E., and Ruzyllo, J. (2007). Bright and color-saturated emission from blue light emitting diodes based on solution processed colloidal nanocrystal quantum dots. *Nano Letters*, **7**(12), 3803.

Tang, T. and Saeedkia, D. (Eds.). (2016). *Advances in Imaging and Sensing*. CRC Press.

Tong, Q.-Y. and Gosele, U. (1999). *Semiconductor Wafer Bonding*. J. Wiley and Sons.

Torek, K., Ruzyllo, J., Grant, R., and Novak, R. (1985). Reduced pressure etching of thermal oxides in anhydrous HF/alcoholic gas mixture. *Journal of the Electrochemical Society*, **142**, 1322.

Tyagi, M. S. (1991). *Introduction to Semiconductor Materials and Devices*. J. Wiley and Sons.

Van Zandt, P. (2014). *Microchip Fabrication: A Practical Guide to Semiconductor Processing* (6th ed.). McGraw-Hill.

Vereecke, G., Rohr, E., and Heyns, M. M. (1999). Laser-assisted removal of particles on silicon wafers. *Journal of Applied Physics*, **85**, 3837.

Vossen, J. L. and Kern, W. (Eds.). (2012). *Thin Film Processes*. Academic Press.

Wakil, A., Bhatia, D., Liu, L., Ali, A., Drummond, P., Datta, S., and Ruzyllo, J. (2012). Photoconductance decay characterization of 3D semiconductor material systems. Electronic Materials Conference.

Wali, F., Knotter, D. M., and Kuper, F. G. (2009). Impact of particles in ultra-pure water on the random yield loss in IC production. *Microelectronic Engineering*, **86**, 140.

Wetzig, K. and Schneider, C. M. (Eds.). (2003). *Metal Based Thin Films for Electronics*. Wiley-VCH.

Wolf, M. F. (1969). *Silicon Semiconductor Data*. Pergamon Press.

Wolf, S. and Tauber, R. (1986). *Silicon Processing for the VLSI Era: Vol. 1 – Process Technology*. Lattice Press.

Wolfe, C. M., Holonyak, N., Jr., and Stillman, G. E. (1989). *Physical Properties of Semiconductors*. Prentice Hall.

Wong, W. S. and Salleo, A. (2009). *Flexible Electronics: Materials and Applications*. Springer-Verlag.

Wu, C.-T., Ridley, R., Roman, P., Dolny, G., Grebs, T., Hao, J., and Ruzyllo, J. (2001). The effect of surface treatments and growth conditions on electrical characteristics of thick (>50 nm) gate oxides. *Journal of the Electrochemical Society*, **148**, F184.

Yacobi, B. G. (2008). *Semiconductor Materials: An Introduction to Basic Principles*. Springer-Verlag.

Yoo, C. S. (2008). *Semiconductor Manufacturing Technology*. World Scientific.

Index

www.ingramcontent.com/pod-product-compliance
Lightning Source LLC
La Vergne TN
LVHW050507201026
850618LV00007B/63
* 9 7 8 9 8 1 9 8 2 6 0 3 2 *